OF

FLORIDA,
GEORGIA
&
ALABAMA

IAN SHELDON

LONE
PINE

THE PUBLISHER: LONE PINE PUBLISHING

1808 B Street NW, Suite 140   10145-81 Avenue
Auburn, WA 98001   Edmonton, AB T6E 1W9
USA   Canada

Website: http://www.lonepinepublishing.com

National Library of Canada Cataloguing in Publication Data

Sheldon, Ian, (date)
    Animal tracks of Florida, Georgia & Alabama

    Includes bibliographical references and index.
    ISBN-10: 1-55105-147-8
    ISBN-13: 978-1-55105-147-5

    1. Animal tracks—Florida—Identification. 2. Animal tracks—
Georgia—Identification. 3. Animal tracks—Alabama—Identification. I. Title.
    QL768.S522 1998   591.47'9   C98-910841-4

Editorial Director: Nancy Foulds
Editorial: Volker Bodegom, Randy Williams, Lee Craig
Production Manager: Jody Reekie
Design, Layout and Production: Volker Bodegom, Monica Triska,
    Arlana Anderson-Hale
Additional Research: Tamara Eder
Technical Review: Donald L. Pattie
Technical Contributor: Mark Elbroch
Animal Illustrations: by Gary Ross, except for those by Ted Nordhagen
    (pp. 107 & 117) and Ewa Pluciennik (p. 113)
Track Illustrations: Ian Sheldon
Cover Design: Robert Weidemann
Cover Illustration: River Otter by Gary Ross
Scanning: Elite Lithographers Ltd.

We acknowledge the financial support of the Government of Canada through
the Book Publishing Industry Development Program (BPIDP)
for our publishing activities.

PC: 6

CONTENTS

INTRODUCTION

If you have ever spent time with an experienced tracker, or perhaps a veteran hunter, then you know just how much there is to learn about the subject of tracking and just how exciting the challenge of tracking animals can be. Maybe you think that tracking is no fun, because all you get to see is the animal's prints. What about the animal itself—is that not much more exciting? Well, for most of us who do not spend a great deal of time in the beautiful wilderness of Florida, Georgia or Alabama, the chances of seeing the fun-loving River Otter or the mysterious Mountain Lion are slim. The closest that we may ever get to some animals will be through their tracks, and they can inspire a very intimate experience. Remember, you are following in the footsteps of the unseen—animals that are in pursuit of prey, or perhaps being pursued *as* prey.

This book offers an introduction to the complex world of tracking animals. Sometimes tracking is easy. At other times it is an incredible challenge that leaves you wondering just what animal left those unusual tracks. Take this book into the field with you, and it can provide some help

Eastern Tiger Salamander

with the first steps to identification. Animals tracks and trails are this book's focus; you will learn to recognize subtle differences for both. There are, of course, many additional signs to consider, such as scat and food caches, all of which help you to understand the animal that you are tracking.

Raccoon

It takes many years to become an expert tracker. Tracking is one of those skills that grows with you as you acquire new knowledge in new situations. Most importantly, you will have an intimate experience with nature. You will learn the secrets of the seldom seen. The more you discover, the more you will want to know. And, by developing a good understanding of tracking, you will gain an excellent appreciation of the intricacies and delights of our marvelous natural world.

How to Use This Book

Most importantly, take this book into the field with you! Relying on your memory is not an adequate way to identify tracks. Track identification has to be done in the field, or with detailed sketches and notes that you can take home. Much of the process of identification involves circumstantial evidence, so you will have much more success when standing beside the track.

This book is laid out in an easy-to-use format. There is a quick reference appendix to the tracks of all the animals in the book that begins on p. 134. The appendix is a fast

Bullfrog

way to familiarize yourself with certain tracks, and it guides you to the more informative descriptions of each animal and its tracks.

Each description is illustrated with the appropriate footprints and the track patterns that it usually leaves. Although these illustrations are not exhaustive, they do show the tracks or groups of prints that you will most likely see. You will find a list of dimensions for the tracks, giving the general range, but there will always be extremes, just as there are with people who have unusually small or large feet. Under the category 'Size' (of animal), the 'greater-than' sign (>) is used when the size difference between the sexes is pronounced.

If you think that you may have identified a track, check the 'Similar Species' section. This section is designed to help you confirm your conclusions by pointing out other

Nutria

Eastern Woodrat

animals that leave similar tracks and showing you ways to distinguish among them.

As you read this book, you will notice an abundance of words such as 'often,' 'mostly' and 'usually.' Unfortunately, tracking will never be an exact science; we cannot expect animals to conform to our expectations, so be prepared for the unpredictable.

Tips on Tracking

You will notice clear, well-formed prints as you flip through this guide. Do not be deceived! It is a rare track that will ever show so clearly. For a good, clear print, the perfect conditions are wet sand, light dust or mud that is slightly soft and not too sloppy. These conditions can be rare—most often you will be dealing with incomplete or faint prints where you cannot even be sure of the number of toes.

Should you find yourself looking at a clear print, then the job of identification is much easier. There are a number of key features to look for: measure the length and width of the print, count the number of toes, check for claw marks and note how far away they are from the body of the print, and look for a heel mark. Keep in mind other, more subtle features, such as the spacing between the toes, whether they are parallel or not, and whether fur on the sole of the foot has made the print less clear.

When you are faced with the challenge of identifying an unclear print—or even if you think that you have made a successful identification from one print alone—look beyond the single footprint and search out others. Do not rely on the dimensions of one print alone, but collect measurements from several prints to get an average impression. Even the prints within one trail can show a lot of variation.

Try to determine which is the fore print and which is the hind, and remember that many animals are built very differently from humans, having larger forefeet than hind feet. Sometimes the prints will overlap, or they can be directly on top of one another in a direct register. For some animals, the fore and hind prints are pretty much the same.

Check out the pattern that the prints make together in the trail, and follow the trail for as many paces as is necessary for you to become familiar with the pattern. Patterns are very important, and they can be the distinguishing feature between different animals with otherwise similar tracks.

Follow the trail for some distance—it can give you some vital clues. For example, the trail may lead you to a tree, indicating that the animal is a climber, or it may lead down into a burrow. This part of tracking can be the most rewarding, because you are following the life of the animal as it hunts, runs, walks, jumps, feeds or tries to escape a predator.

Take into consideration the habitat. Sometimes very similar species can be distinguished only by their locations—one might be found on the riverbank, whereas another might be encountered only in the dense forest.

Think about your geographical location, too, because some animals have a limited range. This consideration can rule out some species and help you with your identification.

Remember that every animal will at some point leave a print or track that looks just like the print or track of a completely different animal!

Finally, keep in mind that if you track quietly, you might catch up with the maker of the prints.

Striped Skunk

Terms & Measurements

Some of the terms used in tracking can be rather confusing, and they often depend on personal interpretation. For example, what comes to your mind if you see the word 'hopping'? Perhaps you see a person hopping along about on one leg, or perhaps you see a rabbit hopping along through the countryside. Clearly, one person's perception of motion can be very different from another's. Some useful terms are explained on the next few pages, to clarify what is meant in this book, and, where appropriate, how the measurements given fit in with each term.

The following terms are sometimes used loosely and interchangeably—for example, a rabbit might be described as 'a hopper' and a squirrel as 'a bounder,' yet both leave the same pattern of prints in the same sequence.

Ambling: Fast, rolling walking.

Bounding: A gait of four-legged animals in which the two hind feet land simultaneously, usually registering in front of the fore prints. It is common in rodents and the rabbit family. 'Hopping' or 'jumping' can often be substituted.

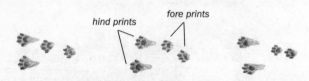

hind prints   fore prints

Gait: An animal's gait describes how it is moving at some point in time. Different gaits result in different observable trail characteristics.

Galloping: A gait used by animals with four legs of even length, such as dogs, moving at high speed, hind feet registering in front of forefeet.

Hopping: Similar to bounding. With four-legged animals, it is usually indicated by tight clusters of prints, fore prints set between and behind the hind prints. A bird hopping on two feet creates a series of paired tracks along its trail.

Loping: Like galloping but slower, with each foot falling independently and leaving a trail pattern that consists of groups of tracks in the sequence fore-hind-fore-hind, usually roughly in a line.

Mustelids (weasel family) often use ***2×2 loping***, in which the hind feet register directly on the fore prints. The resulting pattern has angled, paired tracks.

Running: Like galloping, but applied generally to animals moving at high speed. Also used for two-legged animals.

Trotting: Faster than walking, slower than running. The diagonally opposite limbs move simultaneously; that is, the right forefoot with the left hind, then the left forefoot with the right hind. This gait is the natural one for canids (dog family), short-tailed shrews and voles.

hind print   fore print

Canids may use ***side-trotting***, a fast trotting in which the hind end of the animal shifts to one side. The resulting track pattern has paired tracks, with all the fore prints on one side and all the hind prints on the other.

Walking: A slow gait in which each foot moves independently of the others, resulting in an alternating track pattern. This gait is common for felines (cat family) and deer, as well as wide-bodied animals, such as bears and porcupines. The term is also used for two-legged animals.

Other Tracking Terms

Dewclaws: Two small, toe-like structures set above and behind the main foot of most hoofed animals.

> dewclaw marks

Direct Register: The hind foot falls directly on the fore print.

double register   direct register

Double Register: The hind foot registers and overlaps the fore print only slightly or falls beside it, so that both prints can be seen at least in part.

Dragline: A line left in soft substrates by a foot or the tail dragging over the surface.

dragline

Gallop Group: A track pattern of four prints made at a gallop, usually with the hind feet registering in front of the forefeet (see '**galloping**' for illustration).

Height: Taken at the animal's shoulder.

Length: The animal's body length from head to rump, not including the tail, unless otherwise indicated.

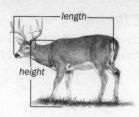

Metacarpal Pad: A small pad near the palm pad—or between the palm pad and heel—on the forefeet of bears and members of the weasel family.

Print (also called '*track*'): Fore and hind prints are treated individually. Print dimensions given are 'length' (including claws—maximum values may represent occasional heel register for some animals) and 'width.' A group of prints made by each of the animal's feet makes up a track pattern.

Register: To leave a mark—said about a foot, claw or other part of an animal's body.

Nine-banded Armadillo

Retractable: Describes claws that can be pulled in to keep them sharp, as with the cat family; these claws do not register in the prints. Foxes have semi-retractable claws.

Sitzmark: The mark left on the ground by an animal falling or jumping from a tree.

Straddle: The total width of the trail, all prints considered.

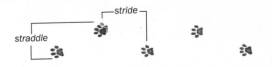

Stride: For consistency among different animals, the stride is taken as the distance from the center of one print (or print group) to the center of the next one. Some books may use the term 'pace.'

Track: Same as '***print***.'

Track or Pattern: The pattern left after each foot registers once, or a set of prints, such as a gallop group.

Trail: A series of track patterns. Think of it as the path of the animal.

MAMMALS

River Otter

White-tailed Deer

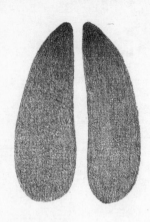

Fore and Hind Prints
Length: 2–3.5 in (5–9 cm)
Width: 1.6–2.5 in (4–6.5 cm)

Straddle
5–10 in (13–25 cm)

Stride
Walking: 10–20 in (25–50 cm)
Galloping: 6–15 ft (1.8–4.5 m)

Size (buck>doe)
Height: 3–3.5 ft (90–110 cm)
Length: to 6.3 ft (1.9 m)

Weight
120–350 lb (55–160 kg)

walking   *gallop group*

WHITE-TAILED DEER
Odocoileus virginianus

The keen hearing of this deer guarantees that it knows about you before you know about it. Frequently, all that we see is its conspicuous white tail as it gallops away, earning this deer the nickname 'flagtail.' The White-tailed Deer, widespread throughout the three states, may be found in small groups at the edges of woodlands. It often ventures out into open areas and is frequently seen in farmland and close to residential areas.

This deer's prints are heart-shaped and pointed. Its alternating walking track pattern shows the hind prints direct registered or double registered on the fore prints. In deep substrate, such as wet sand or mud, the dewclaws register. This flighty deer gallops in the usual style, leaving hind prints in front of fore prints, with toes spread wide for steadier, safer footing.

Similar Species: The White-tailed Deer is the only wild deer in the three states. The Feral Pig (p. 22) usually registers its prominent dewclaws.

Feral Pig

Fore and Hind Prints

(with dewclaws)
Length: 2.5–3 in (6.5–7.5 cm)
Width: 2.3 in (5.8 cm)

Straddle
5–6 in (13–15 cm)

Stride
Trotting: 16–20 in (40–50 cm)

Size (male>female)
Height: 3 ft (90 cm)
Length: 4.3–6 ft (1.3–1.8 m)

Weight
77–440 lb (35–200 kg)

trotting

FERAL PIG (Wild Pig, Wild Boar)

Sus scrofa

Descended from animals introduced by European settlers, the wild Feral Pig interbred with escaped Domestic Pigs. Populations of this sturdy beast of the dense undergrowth can be found scattered through some parts of the region, most notably in Georgia. Armed with tusks, this pig can be quite threatening.

A Feral Pig's print shows two prominent, widely spaced toe marks, and usually, except on firm surfaces, a clear, pointed dewclaw mark off to the side. The hind print is slightly smaller than the fore print. Feral Pigs are keen foragers, so their tracks can often be numerous, especially when they travel in a group. They usually trot, generally making an alternating track pattern typical of four-legged animals, with a double register of hind over fore print. Other signs of the Feral Pig are wallows and diggings.

Similar Species: White-tailed Deer (p. 20) tracks are similar, but more pointed, with a longer stride, dewclaws to the rear (not the side) and a narrower gap between the toes. Although the Domestic Pig is the same species, its tracks have a wider straddle and are less neat, often forming two separate lines.

Horse

Fore Print
(hind print is slightly smaller)
Length: 4.5–6 in (11–15 cm)
Width: 4.5–5.5 in (11–14 cm)

Straddle
2–7.5 in (5–19 cm)

Stride
Walking: 17–27 in (43–70 cm)

Size
Height: to 6 ft (1.8 m)

Weight
to 1500 lb (680 kg)

walking

HORSE
Equus caballus

This popular animal has unmistakable prints. It deserves mention because widespread use of the Horse means that you can expect its tracks to show up almost anywhere.

Unlike any other animal in this book, the Horse has just one huge toe on each foot. This toe leaves an oval print with a distinctive 'frog' (V-shaped mark) at its base. If a Horse is shod, the horseshoe shows up clearly as a firm wall at the outside of the print. Not all horses are shod, so do not expect to see this outer wall on every horse track. A typical, unhurried horse track shows an alternating walk with hind prints registering on or behind the slightly larger fore prints. Horses are capable of a range of speeds—up to a full gallop—but most recreational horseback riders prefer to walk their horses and soak up the local views!

Similar Species: Mules (rarely shod) have smaller tracks.

Black Bear

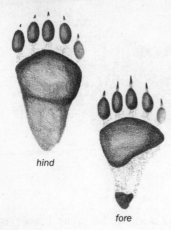

hind

fore

Fore Print
Length: 4–6.3 in (10–16 cm)
Width: 3.8–5.5 in (9.5–14 cm)

Hind Print
Length: 6–7 in (15–18 cm)
Width: 3.5–5.5 in (9–14 cm)

Straddle
9–15 in (23–38 cm)

Stride
Walking: 17–23 in (43–58 cm)

Size (male>female)
Height: 3–3.5 ft (90–110 cm)
Length: 5–6 ft (1.5–1.8 m)

Weight
200–600 lb (90–270 kg)

*walking
(slow)*

BLACK BEAR
Ursus americanus

The Black Bear inhabits the southernmost part of the region, where it prefers wild terrain in forested, swampy areas, and the mountains of northern Georgia, where it may sleep deeply through the colder months. Finding fresh bear tracks can be a thrill, but take care—the bear may be just around the corner. Never underestimate the potential power of a surprised bear!

Black Bear prints somewhat resemble small human prints, but they are wider and show claw marks. The small inner toe rarely registers. The forefoot's small heel pad often shows and the hind foot has a big heel. The bear's slow walk results in a slightly pigeon-toed double register with the hind print on the fore print. More frequently, at a faster pace, the hind foot oversteps the forefoot. When a bear runs, the two hind feet register in front of the forefeet in an extended cluster. Along well-worn bear paths, look for 'digs'—patches of dug-up earth—and 'bear trees' whose scratched bark shows that these bears climb.

Similar Species: No other animal in the region leaves clawed prints this big.

Domestic Dog

fore

hind

Fore Print
(hind print is smaller)
Length: 1–5.5 in (2.5–14 cm)
Width: 1–5 in (2.5–13 cm)

Straddle
1.5–8 in (3.8–20 cm)

Stride
Walking: 3–32 in (7.5–80 cm)
Loping/Galloping: to 9 ft (2.7 m)

Size
Very variable

Weight
Very variable

slow trotting

loping to galloping

DOMESTIC DOG
Canis familiaris

Dogs come in many shapes and sizes, from the tiny Chihuahua with its dainty feet to the robust and powerful Great Dane. Consequently, Domestic Dog tracks vary enormously. Dog ownership is high in many residential areas, and the popular pastime of dog walking can result in many dog tracks being left scattered about, especially in wet mud or sand.

The forefeet of the Domestic Dog, which are much larger than its hind feet and support more of the animal's weight, leave the clearest tracks. When a dog walks, the hind prints usually register ahead of or beside the fore prints. As the dog moves faster, it trots and then lopes before it gallops. In a trot or lope pattern the prints alternate fore-hind-fore-hind, whereas a gallop group shows (from back to front) fore-fore-hind-hind.

Similar Species: Dog tracks are usually found close to human tracks or activity. Red Wolf (*C. rufus*) or Coyote (*C. latrans*) tracks may look like a large Domestic Dog's, but they are very rare in the region. Fox (pp. 30–33) tracks may be confused with a small dog's.

Red Fox

fore

hind

Fore Print
(hind print is slightly smaller)
Length: 2.1–3 in (5.3–7.5 cm)
Width: 1.6–2.3 in (4–5.8 cm)

Straddle
2–3.5 in (5–9 cm)

Stride
Walking: 12–18 in (30–45 cm)
Side-trotting:
   14–21 in (35–53 cm)

Size
(vixen is slightly smaller)
Height: 14 in (35 cm)
Length: 22–25 in (55–65 cm)

Weight
7–15 lb (3.2–7 kg)

walking

side-trotting

RED FOX
Vulpes vulpes

    This beautiful and notoriously cunning fox is found throughout Alabama and into Georgia and northwestern Florida. It prefers meadows and other open areas, streambanks and woodlands. The secretive and largely nocturnal nature of the adaptable and intelligent Red Fox means that sightings are unlikely, even when tracks are abundant.

    Abundant foot hair prevents fine details from registering—just parts of the toes and heel pads show. The horizontal or slightly curved bar across the fore heel pad is diagnostic. A trotting Red Fox leaves a straight trail of alternating prints—the hind print registers directly on the wider fore print. When a fox side-trots, it leaves print pairs in which the hind print falls to one side of the fore print in typical canid fashion. It gallops like the Domestic Dog (p. 28); the faster the gallop, the straighter the gallop group.

Similar Species: Other canid prints lack the bar across the fore heel pad. Gray Fox (p. 32) prints are smaller. Domestic Dog prints can be of similar size, but with a shorter stride and a less direct trail. When a Red Fox's claws do not register, its track can resemble a Domestic Cat's (p. 38).

31

Gray Fox

fore

hind

Fore Print
(hind print is slightly smaller)
Length: 1.3–2.1 in (3.3–5.3 cm)
Width: 1.1–1.5 in (2.8–3.8 cm)

Straddle
2–4 in (5–10 cm)

Stride
Walking/Trotting: 7–12 in (25–30 cm)

Size
Height: 14 in (35 cm)
Length: 21–30 in (53–75 cm)

Weight
7–15 lb (3.2–7 kg)

walking

GRAY FOX
Urocyon cinereoargenteus

Found throughout all three states, this small, shy fox is nocturnal in nature. It can sometimes be seen in abandoned fields and farmland that border the woodland and shrubby areas that are its preferred habitat. The only fox to climb trees, it does so either for safety or to forage. The Gray Fox often dens among rocks or in a hollow tree.

The forefoot registers better than the smaller hind foot, and the hind foot's long, semi-retractable claws do not always register. The heel pad marks are often indistinct—sometimes they show up as just small, round dots. When this fox walks, it leaves a neat alternating pattern. When it trots, its prints fall in pairs, with the fore print set behind the rear print diagonally. The Gray Fox's gallop group is like the Domestic Dog's (p. 28).

Similar Species: The Red Fox (p. 30) has heel pads with a bar across them; in general, its prints are larger and less clear (because of thick fur), its stride is longer, and its straddle is narrower. Domestic Cat (p. 38) and Bobcat (p. 36) prints lack claw marks and show larger, less symmetrical heel pads.

Mountain Lion

fore

hind

Fore Print
(hind print is slightly smaller)
Length: 3–4.3 in (7.5–11 cm)
Width: 3.3–4.8 in (8.5–12 cm)

Straddle
8–12 in (20–30 cm)

Stride
Walking: 13–32 in (33–80 cm)
Bounding: to 12 ft (3.7 m)

Size
Height: 25–32 in (65–80 cm)
Length: 3.5–5 ft (1.1–1.5 m)

Weight
70–200 lb (32–90 kg)

walking (fast)

MOUNTAIN LION
(Puma, Cougar, Panther)
Puma concolor

The magnificent Mountain Lion is shy and nocturnal in nature, and each cat requires a big home territory, so finding its tracks is usually the best that trackers can hope for. It is desperately holding onto its last regional stronghold in the swamps and inaccessible parts of Florida.

Mountain Lion prints tend to be wider than they are long. The retractable claws never register. In winter, thick fur on the foot enlarges the print and may stop the two lobes on the front of the heel pad from registering clearly. When a Mountain Lion walks, the hind print either direct registers or double registers on the larger fore print. As the walking pace increases, the hind print tends to fall ahead of the fore print. In dust or sand, the thick, long tail may leave a dragline that can blur some footprint detail. A Mountain Lion seldom gallops, but when it needs to catch prey, it is capable of long bounds. Also look for partly buried scat and kills covered for later eating.

Similar Species: A large Bobcat's (p. 36) prints may be confused with a juvenile Mountain Lion's.

Bobcat

fore

hind

**Fore Print
(hind print is slightly smaller)**
Length: 1.8–2.5 in (4.5–6.5 cm)
Width: 1.8–2.5 in (4.5–6.5 cm)

Straddle
4–7 in (10–18 cm)

Stride
Walking: 8–16 in (20–40 cm)
Running: 4–8 ft (1.2–2.4 m)

**Size
(female is slightly smaller)**
Height: 20–22 in (50–55 cm)
Length: 25–30 in (65–75 cm)

Weight
15–35 lb (7–16 kg)

walking

*ambling
to loping*

BOBCAT (Wildcat)
Lynx rufus

The Bobcat, a stealthy and usually nocturnal hunter that is widely distributed in Florida and in southern parts of Georgia and Alabama, is seldom seen. Very adaptable, it can leave tracks anywhere from wild mountainsides to chaparral and even in residential areas.

A walking Bobcat's hind feet usually register directly on its larger fore prints. As the Bobcat picks up speed, its trail becomes an ambling pattern of paired prints, the hind leading the fore. At even greater speeds, it leaves four-print groups in a lope pattern. The fore prints show particular asymmetry. The front part of the heel pad has two lobes and the rear part has three. In deep sand or mud, the Bobcat's feet may leave draglines. The Bobcat marks its territory with half-buried scat along its meandering trail.

Similar Species: Juvenile Mountain Lion (p. 34) prints can be similar. Large Domestic Cats (p. 38) have similar prints but a shorter stride and a narrower straddle, and they will not wander far from home. Canid (pp. 28–33) prints are narrower than they are long and show claw marks, and the fronts of their footpads are once-lobed.

Domestic Cat

fore

hind

Fore Print
(hind print is slightly smaller)
Length: 1–1.6 in (2.5–4 cm)
Width: 1–1.8 in (2.5–4.5 cm)

Straddle
2.4–4.5 in (6–11 cm)

Stride
Walking: 5–8 in (13–20 cm)
Loping/Galloping:
    14–32 in (35–80 cm)

Size (male>female)
Height: 20–22 in (50–55 cm)
Length with tail: 30 in (75 cm)

Weight
6.5–13 lb (3–6 kg)

walking

loping to galloping

DOMESTIC CAT
(House Cat)
Felis catus

The tracks of the familiar and abundant Domestic Cat can show up almost any place where there are people. Abandoned cats may roam farther afield; these 'feral cats' lead a pretty wild and independent existence. Domestic Cats can come in many shapes, sizes and colors.

As with all felines, a Domestic Cat's fore print and slightly smaller hind print both show four toe pads. Its retractable claws, kept clean and sharp for catching prey, do not register. Cat prints usually show a slight asymmetry, with one toe leading the others. A Domestic Cat makes a neat alternating walking track pattern, usually in direct register, as one would expect from this animal's fastidious nature. When a cat picks up speed, it leaves clusters of four prints, the hind feet registering in front of the forefeet.

Similar Species: A small Bobcat (p. 36) may leave tracks similar to a very large Domestic Cat's. Domestic Dog (p. 28) and fox (pp. 30–33) prints show claw marks.

Raccoon

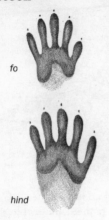

fo

hind

Fore Print
Length: 2–3 in (5–7.5 cm)
Width: 1.8–2.5 in (4.5–6.5 cm)

Hind Print
Length: 2.4–3.8 in (6–9.5 cm)
Width: 2–2.5 in (5–6.5 cm)

Straddle
3.3–6 in (8.5–15 cm)

Stride
Walking: 8–18 in (20–45 cm)
Bounding: 15–25 in (38–65 cm)

Size
(female is slightly smaller)
Length: 24–37 in (60–95 cm)

Weight
11–35 lb (5–16 kg)

walking

bounding
group

RACCOON
Procyon lotor

The inquisitive Raccoon, common throughout all three states, is adored by some people for its distinctive face mask, yet disliked for its curiosity—often demonstrated with residential garbage cans. A good place to look for its tracks is near water. The Raccoon likes to rest in trees. It usually dens up in cold weather.

The Raccoon's unusual print, showing five well-formed toes, looks like a human handprint; its small claws make dots. Its highly dexterous forefeet rarely leave heel prints, but its hind prints, which are generally much clearer, do show heels. The Raccoon's peculiar walking track pattern shows the left fore print next to the right hind print (or just in front) and vice versa. When a raccoon walks in deep substrate, such as mud, it may use a direct-registering walk. The Raccoon occasionally bounds, leaving clusters with the two hind prints in front of the fore prints.

Similar Species: Unclear Opossum (p. 42) prints may look similar, but the Opossum drags its tail and has a distinctive thumb. In snow, also check the River Otter (p. 48).

41

Opossum

fore

hind

Fore Print
Length: 2–2.3 in (5–5.8 cm)
Width: 2–2.3 in (5–5.8 cm)

Hind Print
Length: 2.5–3 in (6.5–7.5 cm)
Width: 2–3 in (5–7.5 cm)

Straddle
4–5 in (10–13 cm)

Stride
5–11 in (13–28 cm)

Size
Length: 2–2.5 ft (60–75 cm)

Weight
9–13 lb (4–6 kg)

walking

fast walking

OPOSSUM
Didelphis virginiana

   This slow-moving,
nocturnal marsupial is found
throughout all three states. Though it
occupies many habitats and is quite
tolerant of residential areas, it prefers open
woodland or brushland around waterbodies.
Opossum tracks are commonly seen in mud near
water. If you find some roadkill, which Opossums
like to eat, look for tracks along the roadside—
though many Opossums suffer the same fate as the
carrion they dine on.

   The Opossum is an excellent climber, so its trail may
lead to a tree. It has two walking habits: the common
alternating pattern, with the hind prints registering on the
fore prints, and a Raccoon-like (p. 40) paired-print pattern,
with each hind print next to the opposing fore print. The
very long, inward-pointing thumb of the hind foot does not
make a claw mark. In sand or dust, the dragline of the tail
may be seen as a thin, gently swerving trail.

Similar Species: Prints in which the distinctive thumbs do
not show may be mistaken for a Raccoon's.

Nine-banded Armadillo

fore

hind

Fore Print
Length: 1.5–1.8 in (3.8–4.5 cm)
Width: 1.4–1.7 in (3.6–4.3 cm)

Hind Print
Length: 2–2.5 in (5–6.5 cm)
Width: 1.5–1.8 in (3.8–4.5 cm)

Straddle
2–3 in (5–7.5 cm)

Stride
3 in (7.5 cm)

Size
Length with tail: 24–32 in (60–80 cm)

Weight
8–17 lb (3.6–7.5 kg)

walking in sand

NINE-BANDED ARMADILLO
Dasypus novemcinctus

This comical, well-armored character of Florida, southern Georgia and Alabama constantly snuffles in the dirt in a pig-like manner. If an armadillo is preoccupied with its affairs, you can get surprisingly close to observe it in action. Hassle it too much and you may get a kick from its strong legs, or it may run off in a straight line and right into a tree! It can run quickly if necessary, but often curls up to protect its vulnerable underside.

Armadillo tracks, often found in an alternating walking pattern, are usually most plentiful near dens and burrows— look in loose, sandy soils that make for easy digging and foraging. When clear, a fore print shows four clawed toes and a hind print shows five, but draglines from the tail and armor often blur the tracks. During hot-weather mud baths, the bony armor may leave imprints. The armadillo may swim rivers or walk across on the bottom.

Similar Species: Clear armadillo tracks are very birdlike. Unclear tracks in sand or dust may be mistaken for White-tailed Deer (p. 20) prints, so look around for other signs if you are unsure.

Harbor Seal

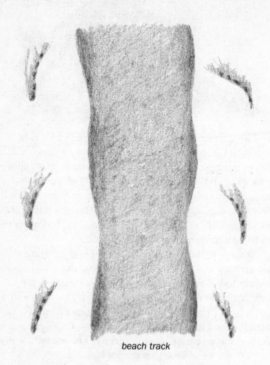

beach track

Size (male>female)
Length: 4–6 ft (1.2–1.8 m)
Weight
180–310 lb (80–140 kg)

HARBOR SEAL
Phoca vitulina

This seal can sometimes be found in Georgia, though it usually frequents isolated beaches farther north. One of the smaller seals, it is quite shy and will usually slide off its rocky sentry post into the sea to escape naturalists. Look in sandy and muddy areas between platforms for its tracks—unmistakable because of their location and large size. Sometimes a Harbor Seal will work its way up a river, so you may also find its tracks along riverbanks.

The seal, though unrivaled in the water, is not the most graceful of animals on land. Its heavy, fat body and flipper-like feet can leave messy tracks: a wide trough (made by its cumbersome belly) with dents alongside (made as the seal pushed itself along with its forefeet). Look for the marks made by the nails.

Similar Species: The Hooded Seal (*Cystophora cristata*) can be an occasional visitor to Florida's coasts but has little reason to come ashore. The West Indian Monk Seal (*Monachus tropicalis*), once seen in the Gulf of Mexico, was too easy a target for early settlers and is now likely extinct.

River Otter

fore

hind

Fore Print
Length: 2.5–3.5 in (6.5–9 cm)
Width: 2–3 in (5–7.5 cm)

Hind Print
Length: 3–4 in (7.5–10 cm)
Width: 2.3–3.3 in (5.8–8.5 cm)

Straddle
4–9 in (10–23 cm)

Stride
Loping: 12–27 in (30–70 cm)

Size
(female is two-thirds the size of male)
Length with tail: 3–4.3 ft (90–130 cm)

Weight
10–25 lb (4.5–11 kg)

loping (fast)

RIVER OTTER
Lontra canadensis

No animal knows how to have more fun than a River Otter. If you are lucky enough to watch one at play, you will not soon forget the experience. Widespread and well-adapted for the aquatic environment, this otter lives along waterbodies in all three states. Expect to find a wealth of evidence along riverbanks in an otter's home territory. An otter in the forest is usually on its way to another waterbody. The River Otter loves to slide down muddy riverbanks, leaving troughs nearly 1 foot (30 cm) wide.

In soft mud, the webbing on the River Otter's five-toed feet, especially the hind ones, may be evident. The inner toes are set slightly apart. If the forefoot's metacarpal pad registers, it lengthens the print. Very variable, otter trails usually show the typical mustelid 2×2 loping. However, with faster gaits, they leave groups of four and three prints. The thick, heavy tail often leaves a dragline.

Similar Species: Although the Mink (p. 50) can leave similar signs (but on a much smaller scale), the size and abundance of the River Otter's signs make identification easy.

Mink

fore

hind

Fore and Hind Prints
Length: 1.3–2 in (3.3–5 cm)
Width: 1.3–1.8 in (3.3–4.5 cm)

Straddle
2.1–3.5 in (5.3–9 cm)

Stride
Walking/Loping: 8–35 in (20–89 cm)

Size
(female is slightly smaller)
Length with tail: 19–28 in (48–70 cm)

Weight
1.5–3.5 lb (0.7–1.6 kg)

2×2 loping

MINK
Mustela vison

The lustrous Mink, widespread throughout the region, prefers watery habitats surrounded by brush or forest. At home as much on land as in water, this nocturnal hunter can be exciting to track. Like the River Otter (p. 48), the Mink likes to slide in mud, carving out a trough up to 6 inches (15 cm) wide for an observant tracker to spot.

The Mink's fore print shows five (perhaps four) toes, and five loosely connected palm pads in an arc, but the hind print shows only four palm pads. The metacarpal pad of the forefoot rarely registers, but the furred heel of the hind foot may register, lengthening the hind print. The Mink prefers the typical mustelid 2×2 loping, making consistently spaced, slightly angled double prints. Its diverse track patterns also include alternating walking, loping with three- and four-print groups (like the River Otter, p. 48) and bounding (like a rabbit, pp. 58–61).

Similar Species: The Long-tailed Weasel (p. 52) makes similar, smaller tracks. The River Otter's webbed feet make much larger prints. Bobcat (p. 36) prints may resemble four-toed Mink prints, but they lack claw marks.

Long-tailed Weasel

Fore and Hind Prints
Length: 1.1–1.8 in (2.8–4.5 cm)
Width: 0.8–1 in (2–2.5 cm)

Straddle
1.8–2.8 in (4.5–7 cm)

Stride
2×2 loping: 9.5–43 in (24–110 cm)

Size (male>female)
Length with tail:
12–22 in (30–55 cm)

Weight
3–12 oz (85–340 g)

2×2 loping

LONG-TAILED WEASEL
Mustela frenata

Weasels are active year-round hunters, with an avid appetite for rodents. The Long-tailed Weasel is found throughout all three states.

Following a weasel's tracks can reveal much about the nimble creature's activities. Tracks are most evident in moist sand or soil after rain. Occasionally, you may find weasel tracks in the sand around a rodent's burrow. Some weasel trails may lead you up a tree. Weasels sometimes take to water.

Because of a weasel's light weight and its small, hairy feet, pad detail is often missing in the tracks, especially in sand or dust. Even on clear prints, the inner (fifth) toe rarely registers. The usual weasel gait is a 2×2 lope that results in a trail of paired prints. The Long-tailed Weasel's typical lope shows an irregular stride with no consistent behavior. Like the Mink (p. 50), this weasel may bound like a rabbit (pp. 58–61).

Similar Species: The Least Weasel (*M. nivalis*), which looks like an extremely small version of the Long-tailed Weasel, can be found in the hills of northern Georgia. Mink tracks are similar, but they are usually wider.

Striped Skunk

fore

hind

Fore Print
Length: 1.5–2.2 in (3.8–5.6 cm)
Width: 1–1.5 in (2.5–3.8 cm)

Hind Print
Length: 1.5–2.5 in (3.8–6.5 cm)
Width: 1–1.5 in (2.5–3.8 cm)

Straddle
2.8–4.5 in (7–11 cm)

Stride
Walking/Bounding:
   2.5–8 in (6.5–20 cm)

Size
Length with tail:
   20–32 in (50–80 cm)

Weight
6–14 lb (2.7–6.5 kg)

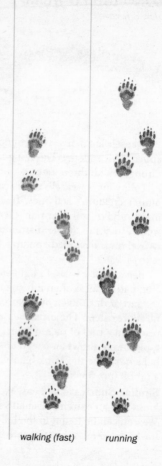

walking (fast)   *running*

STRIPED SKUNK
Mephitis mephitis

This striking skunk has a notorious reputation for its vile smell; the lingering odor is often the best sign of its presence. The Striped Skunk, widespread in all three states, enjoys a diverse range of habitats, especially open woodlands, brushy areas and residential zones. In cold climates it dens up in winter, emerging on warmer days.

Both the fore feet and hind feet have five toes. The long claws on the forefeet often register. Smooth palm pads and small heel pads leave surprisingly small prints. Skunks mostly walk—with such a potent smell for their defense, they rarely need to run. Note that though this skunk's trail rarely shows any consistent pattern, an alternating walking pattern may be evident. The greater the skunk's speed, the more the hind foot oversteps the fore. If it runs, it leaves clumsy, closely set four-print groups. In loose soil or sand it drags its feet.

Similar Species: The Eastern Spotted Skunk (p. 56), with mostly the same range (but missing from southeastern Georgia), leaves smaller prints in a random pattern. Mink (p. 50) and Long-tailed Weasel (p. 52) tracks are spaced farther apart. Skunk prints do not overlap.

Eastern Spotted Skunk

fore

hind

Fore Print
Length: 1–1.3 in (2.5–3.3 cm)
Width: 0.9–1.1 in (2.3–2.8 cm)

Hind Print
Length: 1.2–1.5 in (3–3.8 cm)
Width: 0.9–1.1 in (2.3–2.8 cm)

Straddle
2–3 in (5–7.5 cm)

Stride
Walking: 1.5–3 in (3.8–7.5 cm)
Bounding: 6–12 in (15–30 cm)

Size
Length: 13–25 in (33–65 cm)

Weight
0.6–2.2 lb (0.3–1 kg)

walking   *bounding*

EASTERN SPOTTED SKUNK
Spilogale putorius

   This beautifully marked skunk, smaller than its striped
cousin, is distributed throughout the region, except for
southeastern Georgia. It enjoys diverse habitats—such
as scrubland, forests and farmland—but it is a rare sight,
because of its nocturnal habits, and because it dens up
during cold spells, coming out only on warmer nights.

   This skunk leaves a very haphazard trail as it forages for
food on the ground. Occasionally, and with ease, it climbs
trees. The long claws on the forefeet often register, and
the palm and heel may leave defined pad marks. Although
this skunk rarely runs, when it does so it may bound along,
leaving groups of four prints, hind ahead of fore. It sprays
only when truly provoked, so its powerful odor is less
frequently detected than that of the Striped Skunk (p. 54).

Similar Species: The larger Striped Skunk, found in all
three states, has larger prints and less scattered tracks with a
shorter running stride (or it jumps); it does not climb trees.

Eastern Cottontail

hind

fore

hopping

Fore Print
Length: 1–1.5 in (2.5–3.8 cm)
Width: 0.8–1.3 in (2–3.3 cm)

Hind Print
Length: 3–3.5 in (7.5–9 cm)
Width: 1–1.5 in (2.5–3.8 cm)

Straddle
4–5 in (10–13 cm)

Stride
Hopping: 0.6–3 ft (18–90 cm)

Size
Length: 12–17 in (30–43 cm)

Weight
1.3–3 lb (0.6–1.4 kg)

EASTERN COTTONTAIL
Sylvilagus floridanus

This abundant rabbit is widespread throughout all three states. Preferring brushy areas in grasslands and cultivated areas, it might be found in dense vegetation, hiding from predators such as the Bobcat (p. 36) and foxes (pp. 30–33). Largely nocturnal, the Eastern Cottontail might be seen at dawn or dusk and on darker days.

As with other rabbits and hares, this rabbit's most common track is a triangular grouping of four prints, with the larger hind prints (which can appear pointed) falling in front of the fore prints (which may overlap). The hairiness of the toes will hide any pad detail. If you follow this rabbit's tracks, you could be startled if it flies out from its 'form,' a depression in the ground in which it rests.

Similar Species: The New England Cottontail (*S. transitionalis*), of open areas in northern Georgia and Alabama, leaves similar tracks. The Marsh Rabbit (p. 60) and the Swamp Rabbit (*S. aquaticus*) make similar tracks in wet habitats. The Snowshoe Hare (*Lepus americanus*), with larger prints (especially the hind ones), may be encountered in the mountains. Squirrel (pp. 74–81) tracks show the fore prints more consistently side by side.

Marsh Rabbit

fore

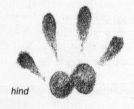

hind

Fore Print
Length: 1–1.5 in (2.5–3.8 cm)
Width: 0.8–1.3 in (2–3.3 cm)

Hind Print
Length: 3–3.5 in (7.5–9 cm)
Width: 1–1.5 in (2.5–3.8 cm)

Straddle
4–5 in (10–13 cm)

Stride
Hopping: 0.6–3 ft (18–90 cm)

Size
Length: 14–18 in (35–45 cm)

Weight
2.4–3.5 lb (1.1–1.6 kg)

walking

hopping

MARSH RABBIT
Sylvilagus palustris

Fond of wet places in southern Georgia and Alabama and in all of Florida, this rabbit can be found anywhere from coastal swamps to inland freshwater lakes. Usually at rest in its 'form' (a depression in the ground) by day, this rabbit comes out after dusk to dine on aquatic vegetation and to dig up the bulbs of its favorite plants.

As is common with rabbits, the Marsh Rabbit has a hopping gait in which the hind feet register ahead of the fore prints. However, unlike most rabbits, the Marsh Rabbit can also walk, leaving an alternating track. Look for its tracks, which may be quite evident in muddy areas, and for well-beaten paths through dense vegetation. Its trail may lead straight into a river, which the rabbit most likely swam across, probably to escape a predator.

Similar Species: Cottontails (p. 58) leave nearly identical tracks. The larger Swamp Rabbit (*S. aquaticus*), in most of Alabama and northwestern Georgia, has a hind print more than 4 inches (10 cm) long. Squirrel (pp. 74–81) tracks have the fore prints more consistently side by side.

61

Nutria

fore

hind

Fore Print
Length: to 3 in (7.5 cm)
Width: to 3 in (7.5 cm)

Hind Print
Length: 4.5–6 in (11–15 cm)
Width: to 3.5 in (9 cm)

Straddle
to 7 in (18 cm)

Stride
Walking: to 8 in (20 cm)

Size (male>female)
Length with tail:
    2.1–4.5 ft (65–140 cm)

Weight
5–25 lb (2.3–11 kg)

walking

NUTRIA
(Swamp Beaver, Coypu)
Myocastor coypus

This large rodent, introduced from South America by fur farmers, has escaped into the wild and now has numerous colonies in the southern states, most notably in southern Alabama, northeastern Florida and the swamps of southern Georgia. Much larger than a Muskrat (p. 66), and with a voracious appetite for rice and sugarcane, the Nutria can be quite destructive to agricultural activities.

The webbing on the Nutria's strong hind feet is often evident in its prints. Look for claw marks, too. The large hind prints each show five toes, one set farther back than the rest. The much smaller fore prints also show five toes. The large, round and hairless tail often leaves a dragline. Nutria tracks may lead you to a den in a riverbank, sometimes a former Muskrat residence. Nearby, you might also spot the large mats of vegetation that are the Nutria's feeding platforms.

Similar Species: The much smaller Muskrat does not have webbed feet. A Beaver (p. 64) has five toes that are more even on each hind foot.

Beaver

hind

Fore Print
Length: 2.5–4 in (6.5–10 cm)
Width: 2–3.5 in (5–9 cm)

Hind Print
Length: 5–7 in (13–18 cm)
Width: 3.3–5.3 in (8.5–13 cm)

Straddle
6–11 in (15–28 cm)

Stride
Walking: 3–6.5 in (7.5–17 cm)

Size
Length with tail: 3–4 ft (90–120 cm)

Weight
28–75 lb (13–34 kg)

walking

BEAVER
Castor canadensis

Few animals leave as many signs of their presence as the Beaver, the largest North American rodent and a common sight around water in Alabama, Georgia and northern Florida. Look for the conspicuous dams and lodges—capable of changing the local landscape—and the stumps of felled trees. Check trunks gnawed clean of bark for marks of the Beaver's huge incisors. A scent mound marked with castoreum, a strong-smelling yellowish fluid which Beavers produce, also indicates recent activity.

The Beaver's thick, scaly tail may mar its tracks, as can the branches that it drags about for construction and food. Check the large hind prints for signs of webbing and broad toenails—the nail of the second inner toe usually does not show. Rarely do all five toes on each foot register. Irregular foot placement in the alternating walking gait may produce a direct register or a double register. Repeated path use results in well-worn trails.

Similar Species: The Beaver's many signs, including large hind prints, minimize confusion. Muskrat (p. 66) prints are smaller.

Muskrat

fore

hind

Fore Print
Length: 1.1–1.5 in (2.8–3.8 cm)
Width: 1.1–1.5 in (2.8–3.8 cm)

Hind Print
Length: 1.6–3.2 in (4–8 cm)
Width: 1.5–2.1 in (3.8–5.3 cm)

Straddle
3–5 in (7.5–13 cm)

Stride
Walking: 3–5 in (7.5–13 cm)
Running: to 1 ft (30 cm)

Size
Length with tail: 16–25 in (40–65 cm)

Weight
2–4 lb (0.9–1.8 kg)

walking

MUSKRAT
Ondatra zibethicus

This rodent is found around water in northern Georgia and in most of Alabama. Beavers are very tolerant of Muskrats and even allow them to live in parts of their lodges. Active all year, the Muskrat leaves plenty of signs. It digs extensive networks of burrows, often undermining riverbanks, so do not be surprised if you suddenly fall into a hidden hole! Also look for small lodges in the water and beds of vegetation on which the Muskrat rests, suns and feeds.

The small fifth (inner) toe of the forefoot rarely registers. Stiff hairs, which aid in swimming, may create a 'shelf' around the five well-formed toes of the hind print. The common walking pattern shows print pairs that alternate from side to side, with the hind print just behind or slightly overlapping the fore print. In snow and sand, a Muskrat's feet drag, and its tail leaves a sweeping dragline.

Similar Species: The smaller Round-tailed Muskrat (p. 68) lives farther south. The Beaver leaves larger prints. Few other animals share this water-loving rodent's habits.

Round-tailed Muskrat

fo

h.

Fore Print
Length: to 1 in (2.5 cm)
Width: to 1 in (2.5 cm)
Hind Print
Length: to 1.5 in (3.8 cm)
Width: to 0.5 in (1.3 cm)
Straddle
2 in (5 cm)
Stride
Walking: 2.5 in (6.5 cm)
Size
Length with tail: 11–15 in (28–38 cm)
Weight
5.5–13 oz (160–370 g)

walking

ROUND-TAILED MUSKRAT
(Florida Water Rat)
Neofiber alleni

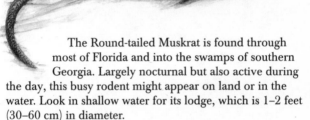

The Round-tailed Muskrat is found through most of Florida and into the swamps of southern Georgia. Largely nocturnal but also active during the day, this busy rodent might appear on land or in the water. Look in shallow water for its lodge, which is 1–2 feet (30–60 cm) in diameter.

The fore print shows four slender toes, and the hind print shows five. The hind print is long and thin, with a heel that usually registers. Look for the marks of the long claws. The tail often drags. This rodent typically walks, leaving an alternating track pattern in which the hind print registers slightly behind the fore print, or it may double register. Repeated use of the same route can create an obvious trail that is about 3 inches (7.5 cm) wide.

Similar Species: The common Muskrat (p. 66) has many similar habits, but it is larger, it has a round tail instead of a vertically flattened one, it has slightly smaller tracks, and it lives farther north.

Woodchuck

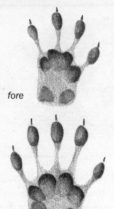

fore

hind

Fore and Hind Prints
Length: 1.8–2.8 in (4.5–7 cm)
Width: 1–2 in (2.5–5 cm)

Straddle
3.3–6 in (8.5–15 cm)

Stride
Walking: 2–6 in (5–15 cm)
Bounding: 6–14 in (15–35 cm)

Size (male>female)
Length with tail:
    20–25 in (50–65 cm)

Weight
5.5–12 lb (2.5–5.5 kg)

walking   *bounding*

WOODCHUCK
(Whistle Pig, Groundhog, Marmot)
Marmota monax

This robust member of the squirrel family is a common sight in open woodlands and cleared areas in northern and central Alabama and in northern Georgia. Always on the watch for predators, but not too troubled by humans, the Woodchuck never roams far from its burrow. This marmot hibernates during winter and emerges in early spring; look for tracks in spring and summer in the dirt around the burrow entrances.

A Woodchuck's fore print shows four toes, three palm pads and two heel pads (not always evident). The hind print shows five toes, four palm pads and two poorly registering heel pads. The Woodchuck usually leaves an alternating walking pattern, with the hind print registered on the fore print. When a Woodchuck runs from danger, it makes groups of four prints, hind ahead of fore.

Similar Species: A small Raccoon's (p. 40) bounding track pattern will be similar, but it will show five-toed fore prints.

Eastern Chipmunk

fore

Fore Print
Length: 0.8–1 in (2–2.5 cm)
Width: 0.4–0.8 in (1–2 cm)

Hind Print
Length: 0.7–1.3 in (1.8–3.3 cm)
Width: 0.5–0.9 in (1.3–2.3 cm)

Straddle
2–3.2 in (5–8 cm)

Stride
Bounding: 7–15 in (18–38 cm)

Size
Length with tail: 7–10 in (18–25 cm)

Weight
2.5–5 oz (70–140 g)

bounding

EASTERN CHIPMUNK
Tamias striatus

This large chipmunk is found in a variety of habitats, from the dense forest floor to open areas near buildings, but it is especially fond of half-cleared forest. Look for this delightful character in northern Georgia and Alabama and into northwestern Florida. You are more likely to see or hear this rodent, which is highly active during summer months, than to notice its tracks. The Eastern Chipmunk is happiest on the ground, but it will gladly climb sturdy oak trees to harvest juicy, ripe acorns. It hibernates lightly in winter, waking up from time to time for a meal and some-times a brief run outside its burrow.

Chipmunks are so light that their tracks rarely show fine details. The forefeet each have four toes, and the hind feet have five. Chipmunks run on their toes, so the two heel pads of the forefeet seldom register; the hind feet have no heel pads. Their erratic track patterns, like those of many of their cousins, show the hind feet registered in front of the forefeet. A chipmunk trail often leads to extensive burrows.

Similar Species: Midwinter tracks were more likely made by a tree squirrel (pp. 74–81), whose tracks will usually be larger. Mouse (pp. 92–95) tracks are smaller.

Eastern Gray Squirrel

fore

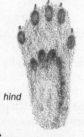

hind

bounding

Fore Print
Length: 1–1.8 in (2.5–4.5 cm)
Width: 1 in (2.5 cm)

Hind Print
Length: 2.3–3 in (5.8–7.5 cm)
Width: 1.1–1.5 in (2.8–3.8 cm)

Straddle
3.8–6 in (9.5–15 cm)

Stride
Bounding: 0.7–3 ft (22–90 cm)

Size
Length with tail: 17–20 in (43–50 cm)

Weight
14–25 oz (400–710 g)

EASTERN GRAY SQUIRREL
Sciurus carolinensis

This large and familiar squirrel can be a common sight in the region's deciduous and mixed forests, even in urban areas. Active all year, the Eastern Gray Squirrel can leave a wealth of evidence, especially in winter as it scurries about digging up nuts that it buried during the previous fall.

The Eastern Gray Squirrel leaves a typical squirrel track when it runs or bounds. The hind prints fall slightly ahead of the fore prints. A clear fore print shows four toes with sharp claws, four fused palm pads and two heel pads. The hind print shows five toes and four palm pads; if the full length of the heel registers, it shows two small heel pads.

Similar Species: The Fox Squirrel (p. 76), with prints as large or larger, occupies the same range. A Red Squirrel (p. 78), which can be found in northern Georgia, has prints that are smaller. A rabbit (pp. 58–61) makes a longer print pattern, and its fore prints rarely register side by side when it runs. Eastern Chipmunk (p. 72) tracks and Southern Flying Squirrel (p. 80) tracks have similar patterns, but they have smaller straddles and prints.

Fox Squirrel

fore

hind

bounding

Fore Print
Length: 1–1.9 in (2.5–4.5 cm)
Width: 1–1.7 in (2.5–4.3 cm)

Hind Print
Length: 2–3.3 in (5–7.5 cm)
Width: 1.5–1.9 in (3.8–4.8 cm)

Straddle
4–6 in (10–15 cm)

Stride
Bounding: 0.7–3 ft (22–90 cm)

Size
Length with tail: 18–28 in (45–70 cm)

Weight
1–2.4 lb (0.5–1.1 kg)

FOX SQUIRREL
Sciurus niger

The Fox Squirrel can be a common sight in all three states, but it has much color variation. While it resembles the Eastern Gray Squirrel (p. 74), it is larger and has a yellowish underside. Look for it in deciduous forests with nut trees, in open areas near woodlots and in Florida's mangroves. Its favorite feeding sites are indicated by piles of nutshells at tree bases. Active all year, this squirrel spends a lot of time foraging on the ground, often collecting nuts that it buried during the previous fall.

When it runs or bounds, the Fox Squirrel makes a typical squirrel track, with the hind prints slightly in front of the fore prints, each pair falling roughly side by side. A clear fore print shows four toes with evident claws, four fused palm pads and two heel pads. The hind print shows five toes, four palm pads and sometimes a heel.

Similar Species: Eastern Gray Squirrel prints are generally slightly smaller. Red Squirrel (p. 78) prints are smaller still. Chipmunk (p. 72) and flying squirrel (p. 80) tracks are in a similar pattern, but they are even smaller and with narrower straddles. A rabbit (pp. 58–61) makes a longer print pattern, and its fore prints rarely register side by side when it runs.

Red Squirrel

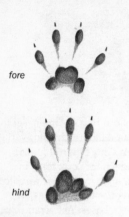

fore

hind

Fore Print
Length: 0.8–1.5 in (2–3.8 cm)
Width: 0.5–1 in (1.3–2.5 cm)

Hind Print
Length: 1.5–2.3 in (3.8–5.8 cm)
Width: 0.8–1.3 in (2–3.3 cm)

Straddle
3–4.5 in (7.5–11 cm)

Stride
Bounding: 8–30 in (20–75 cm)

Size
Length with tail:
    9–15 in (23–38 cm)

Weight
2–9 oz (57–260 g)

bounding

RED SQUIRREL
(Pine Squirrel, Chickaree)
Tamiasciurus hudsonicus

When you enter its territory, a Red Squirrel will greet you with a loud, chattering call. Another obvious sign of this squirrel, which lives in the hills and mountains of northern Georgia, is the large middens—piles of cone scales and cores at the bases of trees—that indicate favorite feeding sites.

Active all year in their small territories, Red Squirrels leave an abundance of trails that lead from tree to tree or down a burrow. These energetic animals mostly bound. This gait leaves groups of four prints, with the hind prints falling in front of the fore prints, which tend to be side by side (but not always). Four toes show on each fore print, and five on each hind print. The heels often do not register when squirrels move quickly.

Similar Species: The Fox Squirrel (p. 76) and the Eastern Gray Squirrel (p. 74), both with a wider range, leave larger tracks. Chipmunks (p. 72) and flying squirrels (p. 80) make smaller tracks with a narrower straddle. When a rabbit (pp. 58–61) hops, it makes longer groups of four prints and its fore prints rarely register side by side.

Southern Flying Squirrel

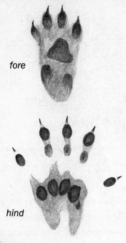

fore

hind

Fore Print
Length: 0.3–0.5 in (0.8–1.3 cm)
Width: 0.4 in (1 cm)

Hind Print
Length: 0.9–1.3 in (2.3–3.3 cm)
Width: 0.5 in (1.4 cm)

Straddle
2–2.5 in (5–6.5 cm)

Stride
Bounding: 7–22 in (18–55 cm)

Size
Length with tail: 8–10 in (20–25 cm)

Weight
1.5–3.2 oz (43–90 g)

bounding

SOUTHERN FLYING SQUIRREL
Glaucomys volans

This soft-furred brown acrobat can be found anywhere from the deciduous mountain forests of the north to the evergreen and gum woods of the south. Primarily nocturnal, the Southern Flying Squirrel is capable of long-distance gliding using the membranes that connect its forelegs and hind legs. In winter, up to 50 Southern Flying Squirrels can be found huddled together in a nest for warmth. Like other tree squirrels, this one does not truly hibernate.

Flying squirrel tracks are hard to find because these animals spend much of their time in trees. This flying squirrel sometimes bounds in a manner similar to the Northern Flying Squirrel, but with a narower straddle that does not exceed 3 inches (7.6 cm). However, information from Mark Elbroch, a tracking expert from the East, shows that the more characteristic pattern for the Southern Flying Squirrel is a bound where the front tracks register in front of the rear tracks, and the straddle is even narrower.

Similar Species: In the mountains of Georgia, the Northern Flying Squirrel (*G. sabrinus*) makes larger prints. Other tree squirrels (pp. 74–79) usually make larger prints and rarely leave sitzmarks. Chipmunks (p. 72) have smaller prints and a narrower straddle.

Eastern Woodrat

fore

hind

Fore Print
Length: 0.6–0.8 in (1.5–2 cm)
Width: 0.4–0.5 in (1–1.3 cm)

Hind Print
Length: 1–1.5 in (2.5–3.8 cm)
Width: 0.6–0.8 in (1.5–2 cm)

Straddle
2.3–2.8 in (5.8–7 cm)

Stride
Walking: 1.8–3 in (4.5–7.5 cm)
Bounding: 5–8 in (13–20 cm)

Size
Length with tail:
   12–18 in (30–45 cm)

Weight
7–16 oz (200–450 g)

walking

bounding

EASTERN WOODRAT
(Florida Woodrat)
Neotoma floridana

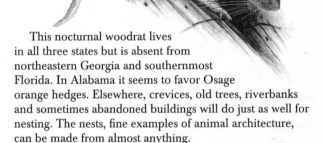

This nocturnal woodrat lives
in all three states but is absent from
northeastern Georgia and southernmost
Florida. In Alabama it seems to favor Osage
orange hedges. Elsewhere, crevices, old trees, riverbanks
and sometimes abandoned buildings will do just as well for
nesting. The nests, fine examples of animal architecture,
can be made from almost anything.

Four toes show on the fore print and five on the hind.
The short claws rarely register. A woodrat often walks in an
alternating fashion, with the hind print direct registering on
the fore print. This woodrat also frequently bounds, leaving
a pattern of four prints, with the larger hind print in front
of the diagonally placed fore prints. The stride tends to be
short relative to the size of the prints. Do not be surprised
if a woodrat trail ends at the base of a tree that it climbed.

Similar Species: The Norway Rat (p. 84) has similar
tracks, but it is usually found close to human activity. The
Woodchuck (p. 70) has similar, much larger prints.

Norway Rat

fore

hind

walking

Fore Print
Length: 0.7–0.8 in (1.8–2 cm)
Width: 0.5–0.7 in (1.3–1.8 cm)

Hind Print
Length: 1–1.3 in (2.5–3.3 cm)
Width: 0.8–1 in (2–2.5 cm)

Straddle
3 in (7.5 cm)

Stride
Walking: 1.5–3.5 in (3.8–9 cm)
Bounding: 9–20 in (23–50 cm)

Size
Length with tail:
    13–19 in (33–48 cm)

Weight
7–18 oz (200–510 g)

NORWAY RAT
(Brown Rat)
Rattus norvegicus

This despised rat is widespread almost anywhere that humans have decided to build homes or operate farms. Not entirely dependent on people, it may be found in the wild as well. It is active both day and night.

The fore print shows four toes and the hind print five. When it bounds, this colonial rat leaves four-print groups, with the hind prints in front of the diagonally placed fore prints. Sometimes one of the hind feet direct registers on a fore print, creating a three-print group. More commonly, rats leave an alternating walking pattern, with the larger hind prints close to or overlapping the fore prints; the hind heel does not show. In dust or sand the tail leaves a dragline. Rats live in groups, so you may find many trails together, often leading to their 5-inch (2-cm) wide burrows.

Similar Species: The Eastern Woodrat's (p. 82) tracks may be similar, but woodrats rarely associate with any human activity, except in abandoned buildings. Mouse (pp. 92–95) prints are much smaller. Squirrel (pp. 74–81) tracks show distinctive squirrel characteristics.

Hispid Cotton Rat

fore

hind

Fore Print
Length: 0.5–0.7 in (1.3–1.8 cm)
Width: 0.5–0.7 in (1.3–1.8 cm)

Hind Print
Length: 0.6–1 in (1.5–2.5 cm)
Width: 0.6–0.8 in (1.5–2 cm)

Straddle
1.3–1.5 in (3.3–3.8 cm)

Stride
Walking: 1.3 in (3.3 cm)

Size
Length with tail: 8–14 in (20–35 cm)

Weight
2.8–7 oz (80–200 g)

walking

HISPID COTTON RAT
Sigmodon hispidus

The Hispid Cotton Rat, widespread in all three states, makes itself unpopular by eating valuable crops. Though keen on eating almost anything green, it prefers grassy fields. It stays close to home, and its little runways clearly mark its routes to favored feeding sites.

The fore prints show four toes, but the larger hind prints usually show five. The heel of the hind foot will not always register, especially if the rat is moving fast. This common medium-sized rodent leaves a typical walking track pattern in which the hind print is on top of and slightly behind the fore print. Watch for its nests, which are balls of woven grass, and for small piles of cut grass.

Similar Species: The widespread Marsh Rice Rat (*Oryzomys palustris*) prefers marshes and leaves slightly smaller tracks. The Key Rice Rat (*O. argentatus*) can be found in the Florida Keys. The Norway Rat (p. 84) makes similar tracks, but it is usually found close to human activity. Eastern Woodrat (p. 82) tracks are similar but with a wider straddle.

Southeastern Pocket Gopher

fore

hind

Fore Print
Length: 1 in (2.5 cm)
Width: 0.6 in (1.5 cm)

Hind Print
Length: 0.8–1.5 in (2–3.8 cm)
Width: 0.5 in (1.3 cm)

Straddle
1.5–2 in (3.8–5 cm)

Stride
Walking: 1.3–2 in (3.3–5 cm)

Size (male>female)
Length with tail: 9–13 in (23–33 cm)

Weight (male)
to 18 oz (510 g)

walking

SOUTHEASTERN POCKET GOPHER

Geomys pinetis

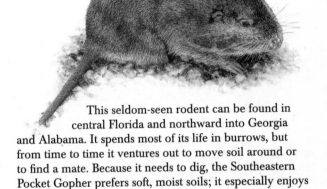

This seldom-seen rodent can be found in central Florida and northward into Georgia and Alabama. It spends most of its life in burrows, but from time to time it ventures out to move soil around or to find a mate. Because it needs to dig, the Southeastern Pocket Gopher prefers soft, moist soils; it especially enjoys pastureland. In coastal regions of Georgia and in Florida the hallmark of this enthusiastic burrower is its bright sandy mounds and tunnel cores. Each mound marks the entrance to a burrow, which is always blocked up with a plug. Search around the mounds to find tracks.

Both fore and hind feet have five toes. Though the forefeet have long, well-developed claws for digging, the tracks rarely show this much detail. Pocket gophers typically walk, making an alternating track pattern in which the hind print registers on or slightly behind the fore print.

Similar Species: A pocket gopher's burrows and mounds may be mistaken for those of moles (p. 98).

Woodland Vole

fore

hind

Fore Print
Length: 0.5 in (1.3 cm)
Width: 0.5 in (1.3 cm)

Hind Print
Length: 0.6 in (1.5 cm)
Width: 0.5–0.8 in (1.3–2 cm)

Straddle
1.3–2 in (3.3–5 cm)

Stride
Walking/Trotting: 0.8 in (2 cm)
Bounding: 2–6 in (5–15 cm)

Size
Length with tail:
    4–5.5 in (10–14 cm)

Weight
0.8–1.3 oz (23–37 g)

walking

WOODLAND VOLE (Pine Vole)

Microtus pinetorum

Though a positive identification of a vole track can be challenging, the most likely candidate throughout Georgia and Alabama and just into northern Florida is the Woodland Vole. It lives in pine forests and in just about every other terrestrial habitat.

When clear (which is seldom), vole fore prints show four toes, and hind prints show five. A vole's walk and trot both leave a paired alternating track pattern with a hind print occasionally direct registered on a fore print. Voles usually opt for a faster bounding; the resulting print pairs show the hind prints registered on the fore prints. These voles lope quickly across open areas, creating a three-print track pattern. In summer, well-used vole paths appear as little runways in the grass. Look for the distinctive piles of cut grass along their runways. The bark at the bases of shrubs may show tiny teeth marks left by gnawing.

Similar Species: The Southern Red-backed Vole (*Clethrionomys gapperi*) and the Rock (Yellownose) Vole (*M. chrotorrhinus*) inhabit the mountains of extreme northern Georgia. The Meadow Vole (*M. pennsylvanicus*) lives in eastern Georgia. Mouse (pp. 92–95) tracks show different patterns.

Cotton Mouse

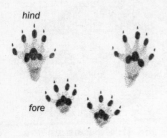

hind

fore

bounding group

Fore Print
Length: 0.3 in (0.8 cm)
Width: 0.3 in (0.8 cm)

Hind Print
Length: 0.6 in (1.5 cm)
Width: 0.4 in (1 cm)

Straddle
1.4–1.8 in (3.6–4.5 cm)

Stride
Bounding: 2–5 in (5–13 cm)

Size
Length with tail:
   6–12 in (15–30 cm)

Weight
0.7–1.6 oz (20–45 g)

bounding

COTTON MOUSE
Peromyscus gossypinus

The nocturnal and seldom-seen Cotton Mouse is widespread throughout Alabama and Florida, and it also inhabits most of Georgia, except the northeast. It prefers swampland but also frequents forests, rocky areas and even beaches. Its tracks may lead you up a tree, down a burrow or into a river—this mouse is a capable swimmer.

A Cotton Mouse fore print shows four toes, three palm pads and two heel pads. A hind print shows five toes (the fifth is often unclear) and three palm pads; the heel pads rarely register. Bounding tracks show hind prints in front of close-set fore prints. In loose dust or sand, the prints may merge and appear as larger pairs, with tail drag evident.

Similar Species: Many other mice have identical tracks. The White-footed Mouse (*P. leucopus*) lives in northern Georgia and Alabama. The Golden Mouse (*Ochrotomys nuttalli*), Oldfield Mouse (*P. polionotus*) and Eastern Harvest Mouse (*Reithrodontomys humulis*) live in all three states, but not in southern Florida. The large-footed Florida Mouse (*Podomys floridanus*) occurs only in Florida. The House Mouse (*Mus musculus*) associates with humans. Jumping mouse (p. 94) prints show long, thin toes. Voles (p. 90) tend to trot and have a longer stride. Shrews (p. 96) have a narrower straddle. A chipmunk's (p. 72) straddle is wider.

Meadow Jumping Mouse

fore

hind

bounding group

Fore Print
Length: 0.3–0.5 in (0.8–1.3 cm)
Width: 0.3–0.5 in (0.8–1.3 cm)

Hind Print
Length: 0.5–1.3 in (1.3–3.3 cm)
Width: 0.5–0.7 in (1.3–1.8 cm)

Straddle
1.8–1.9 in (4.5–4.8 cm)

Stride
Bounding: 7–18 in (18–45 cm)
In alarm: 3–6 ft (90–180 cm)

Size
Length with tail: 7–9 in (18–23 cm)

Weight
0.6–1.3 oz (17–37 g)

bounding

MEADOW JUMPING MOUSE
Zapus hudsonius

Congratulations if you find and successfully identify the tracks of a Meadow Jumping Mouse! This rodent lives in eastern Alabama and into northern Georgia, but it is hard to find. Its preference for grassy meadows—and its winter hibernation—makes locating tracks very difficult.

Jumping mouse tracks are distinctive if you do find them. The two smaller forefeet register between the long hind feet; the long heels do not always register and some prints show just the three long middle toes. The toes on the forefeet may splay so much that the side toes point backward. When bounding, these mice make short leaps. The tail may leave a dragline in soft mud or unseasonable snow. A more abundant sign of this rodent is the clusters of grass stems, cut to about 5 inches (13 cm) long, which are found lying in meadows.

Similar Species: The Woodland Jumping Mouse (*Napaeozapus insignis*) is found in wet, forested areas of northern Georgia. A Cotton Mouse (p. 92) track may have the same straddle. Heel-less hind prints may be mistaken for a vole's (p. 90) or a small bird's (p. 116) or an amphibian's (pp. 118–123).

Southeastern Shrew

hind

fore

running group

Fore Print
Length: 0.2 in (0.5 cm)
Width: 0.2 in (0.5 cm)

Hind Print
Length: 0.6 in (1.5 cm)
Width: 0.3 in (0.8 cm)

Straddle
0.8–1.3 in (2–3.3 cm)

Stride
Bounding: 1.2–2.3 in (3–5.8 cm)

Size
Length with tail: 3–4 in (7.5–10 cm)

Weight
0.1–0.2 oz (3–6 g)

bounding

SOUTHEASTERN SHREW
(Bachman's Shrew)

Sorex longirostris

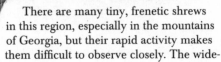

There are many tiny, frenetic shrews in this region, especially in the mountains of Georgia, but their rapid activity makes them difficult to observe closely. The widespread Southeastern Shrew, one of the more likely candidates, can be found in many habitats, especially moist ones. However, it is absent from Florida's south and northwest and from nearby southern Alabama.

In its energetic and unending quest for food, a shrew usually leaves a four-print bounding pattern, but it may slow to an alternating walking gait. The individual prints in a group are often indistinct, but in mud or fine, wet sand you can even count the five toes on each print. In loose dirt a shrew's tail often leaves a dragline. A shrew's trail may disappear down a burrow.

Similar Species: Other shrews have similar tracks. The Southern Short-tailed Shrew (*Blarina carolinensis*) is also widespread. The Northern Short-tailed Shrew (*B. brevicauda*) is found in northern Alabama and Georgia. Slightly smaller, the Least Shrew (*Cryptotis parva*) is found throughout the three states. Many others are confined to the mountains in the north. Mouse (pp. 92–95) fore prints show four toes.

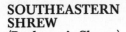

97

Eastern Mole

A molehill of the Eastern Mole

Some molehills and ridges of the Eastern Mole

Size
Length: 4.5–6.5 in (11–17 cm)
Tail length: 1–1.5 in (2.5–3.8 cm)

Weight
2.5–5 oz (70–140 g)

EASTERN MOLE
Scalopus aquaticus

   This soft-furred resident of the underworld is the mole that you will most likely encounter in this region. It can certainly leave a wealth of evidence that indicates its presence, usually in pastures or open woodlands, especially where the soil is light and moist and easy to burrow in.

   Moles, who seldom emerge from their subterranean environment, create an extensive network of burrows through which they forage. These burrows are sometimes marked by ridges on the surface, though most of us are more familiar with the hills that form where the mole gets rid of excess soil from its burrows (and for which moles are frequently considered to be pests when they mess up fine lawns). When rains moisten the soil and bring worms to the surface, it can be entertaining to watch the earth twitch and rise up as the mole satisfies its voracious appetite.

Similar Species: The Star-nosed Mole (*Condylura cristata*), with its strange nose, smaller body and longer tail, prefers wetter areas, often along riverbanks, in Georgia's northern mountains and from southeastern Georgia just into Florida.

BIRDS, AMPHIBIANS & REPTILES

A guide to the animal tracks of these three states is not complete without some consideration of birds, amphibians and reptiles.

Several bird species have been chosen to represent the main types common to this region. Remember, however, that individual bird species are not easily identified by track alone. The shores of lakes and streams are very reliable locations in which to find bird tracks—the mud there can hold a clear print for a long time. The sheer number of tracks made by shorebirds and waterfowl can be truly astonishing. Though some bird species prefer to perch in trees or soar across the sky, it can be entertaining to follow the tracks of birds that spend a lot of time on the ground. They can spin around in circles and lead you in all directions. The trail may suddenly end where the bird took flight, or it might terminate in a pile of feathers, the bird having fallen victim to a hungry predator.

Many amphibians and turtles depend on moist environments, so look in the soft mud along the shores of lakes and ponds for their distinctive tracks. Though you may be able to distinguish frog tracks from toad tracks, because these two amphibians generally move differently, it can be very difficult to identify the species. Reptiles thrive and outnumber the amphibians in drier environments, but they seldom leave good tracks, except in occasional mud or perhaps in sand. Snakes, being without feet, leave distinctive body prints.

Mallard

Print
Length: 2–2.5 in (5–6.5 cm)

Straddle
4 in (10 cm)

Stride
to 4 in (10 cm)

Size
23 in (58 cm)

MALLARD
Anas platyrhynchos

male

female

This dabbling duck is common in open areas by lakes and ponds throughout the three states. The male is an especially familiar sight with its striking green head. The Mallard's prints are often abundant along the muddy shores of most waterbodies, including those in urban parks.

The Mallard's webbed foot has three long toes that all point forward. Though the toes register well, the webbing between them does not always show on the print. Inward-pointing feet give the Mallard a pigeon-toed appearance and may account for its waddling gait, a characteristic for which ducks are known.

Similar Species: Various shorebirds, gulls and dabblers, such as the elaborate Wood Duck (*Aix sponsa*), have similar prints. Exceptionally large webbed prints are likely from a goose, such as the winter-visiting Canada Goose (*Branta canadensis*).

Great Blue Heron

Print
Length: to 6.5 in (17 cm)

Straddle
8 in (20 cm)

Stride
9 in (23 cm)

Size
4.2–4.5 ft (1.3–1.4 m)

GREAT BLUE HERON

Ardea herodias

The refined and graceful image of this large heron symbolizes the precious wetlands in which it patiently hunts for food. Usually still and statuesque as it waits for a meal to swim by, the Great Blue Heron does walk from time to time, perhaps to find a better hunting location. Look for its large, slender tracks along the banks or mud-flats of waterbodies. Not surprisingly, a bird that lives and hunts with such precision walks in a similar fashion, leaving straight tracks that fall in a nearly straight line. Look for the slender rear toe in the print.

Similar Species: Cranes (*Grus* spp.), which occupy similar habitats, have similarly sized prints, but a crane's rear toes are smaller and do not register. Bitterns (*Botaurus* spp.), Ibises (*Plegadis* spp.) and Rails (*Rallus* spp.) are among the waterfowl with similar tracks. Do not be fooled by the Common Moorhen (*Gallinula chloropus*) or the Purple Gallinule (*Pophyrula martinica*)—both have enormous feet (compared to the body) that allow them to stay on top of deep mud or floating vegetation.

Willet

Print
Length: 2–2.5 in (5–6.5 cm)

Straddle
10 in (25 cm)

Stride
Walking: to 9 in (23 cm)

Size
15 in (38 cm)

WILLET
Catoptrophorus semipalmatus

   This common year-round resident of the coastal regions certainly makes its presence known with a loud territorial call of *pill-will-willet*. It takes a good eye, however, to distinguish it from the many other sandpipers on the coast, especially while migrant species are wintering there. Sandpiper prints show only three toes; a very small fourth toe angles off to one side. A sandpiper's stride may be somewhat erratic as it pokes about in quest of food.

Similar Species: Other sandpipers and plovers make similar tracks, though print sizes vary. The common Killdeer (*Charadrius vociferus*) equally enjoys the coast, open fields and riverbanks. This year-round resident's sonorous call repeats its name. Get too close and it might feign a broken wing to distract you from its nearby nest. Killdeer tracks are just over 1 inch (2.5 cm) long.

Bald Eagle

Print
Length: 6 in (15 cm)

Straddle
to 10 in (25 cm)

Stride
Walking: to 9 in (23 cm)

Size
32–37 in (80–95 cm)

BALD EAGLE
Haliaeetus leucocephalus

The magnificent Bald Eagle, with its striking white head and brilliant yellow bill, is a familiar image to most of us. This predator is a scattered but year-round resident of Florida and a winter resident in Georgia and Alabama. Because it feeds largely on fish, it prefers coasts, rivers and lakes. Once persecuted by hunters, this icon of America is making a gradual comeback.

The track of a Bald Eagle is large and robust. Three toes extend forward, and one extends to the rear. The sharp and impressive talons register distinctly. If you find the remnants of fish or other carrion, look closely for Bald Eagle tracks. You will not find many, because this bird rarely spends much time on the ground, preferring vantage points at the tops of trees or cliffs—which is where it also likes to build its huge nests.

Similar Species: The Turkey Vulture (*Cathartes aura*) has similarly sized tracks. Many smaller birds of prey have similar prints in smaller sizes.

Great Horned Owl

Strike
Width: to 3 ft (90 cm)
Size
22 in (55 cm)

GREAT HORNED OWL
Bubo virginianus

Often seen resting quietly in trees by day, this wide-ranging owl prefers to hunt at night. In northern Georgia and Alabama you might find an untidy mark in loose dirt or sandy banks, possibly surrounded by wing and tail-feather imprints. A well-registered 'strike' can be quite a sight. The Great Horned Owl strikes with its talons, and the feather imprints are made as the owl struggles to take off with possibly heavy prey. An ungraceful walker, it prefers to fly away from the scene, though its tracks may be evident near roadkill.

You might stumble across a strike and guess that the owl's target could have been a mouse or vole scurrying around on open ground. Or you may be following the surface trail of an animal to find that it abruptly ends with this strike mark, where the animal has been seized.

Similar Species: If the strike mark has rounded, indistinct feather imprints, it is likely an owl. If not, the strike could be made by a hawk—their feathers are more pointed with well-defined edges.

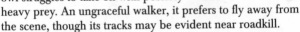

111

Northern Flicker

Print
Length: 1.8 in (4.5 cm)

Straddle
1–1.5 in (2.5–3.8 cm)

Stride
Hopping: 1.5–5 in (3.8–13 cm)

Size
5.5–6.5 in (14–17 cm)

NORTHERN FLICKER

Colaptes auratus

male

female

   This attractive woodpecker, widespread throughout the three states, is common in open woodlands and right into suburban areas. Unlike most woodpeckers, it spends some of its time feeding on the ground.

   A clear flicker track shows a distinctive arrangement of two strong toes pointing forward and two pointing to the rear, with the outer toes slightly longer than the inner ones. The flicker's toes—along with short, strong legs that give the bird a short stride—are well suited for grasping tree trunks and limbs as this agile bird works its way along in search of insects.

Similar Species: Most other birds have very different tracks. Other woodpeckers would leave similar tracks, but very few come down to the ground as much as the obliging Northern Flicker does.

American Crow

Print
Length: 2.5–3 in (6.5–7.5 cm)

Straddle
1.5–3 in (3.8–7.5 cm)

Stride
Walking: 4 in (10 cm)

Size
16 in (40 cm)

AMERICAN CROW
Corvus brachyrhynchos

The black silhouette of the American Crow can be a common sight throughout these three states. A crow will frequently come down to the ground and contentedly strut around with a confidence that hints at its intelligence. Its loud *caw* can be heard from quite a distance. Crows can be especially noisy when they are mobbing an owl or a hawk.

The American Crow typically leaves an alternating walking track pattern. Each print shows three sturdy toes pointing forward and one toe pointing backward. When a crow needs greater speed, perhaps in preparing for take-off, it bounds along, leaving irregular pairs of diagonally placed prints with a longer stride between each successive pair.

Similar Species: Other corvids (members of the crow family), all of which spend a lot of time poking around on the ground, have similar tracks. The Fish Crow (*C. ossifragus*), very similar in appearance but a little smaller, prefers marshes along the coast and in river valleys.

Northern Cardinal

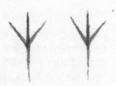

Print
Length to: 1.5 in (3.8 cm)
Straddle
1–1.5 in (2.5–3.8 cm)
Stride
Hopping: 1.5–5 in (3.8–13 cm)
Size
9 in (23 cm)

NORTHERN CARDINAL
Cardinalis cardinalis

female

male

  This common bird typifies the many small hopping birds found in this region. The brilliant red plumage and small black mask and chin of the male are a joy to see on this year-round resident. Each foot has three forward-pointing toes and one longer toe at the rear. The best prints are left in mud, although in really wet mud the toe detail is lost and the feet may show some dragging between the hops.

  A good place to study these types of prints is near a birdfeeder. Watch the birds scurry around as they pick up fallen seeds, then have a look at the prints that they have left behind.

Similar Species: Finches and sparrows make similar tracks. The size of the toes may indicate what kind of bird you are tracking—larger birds have larger footprints. Not all birds are present year-round, so keep the season in mind when tracking.

Frogs

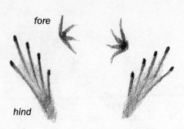

fore

hind

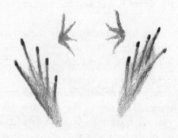

Straddle
to 3 in (7.5 cm)

hopping

118

FROGS

Bullfrog

Look for frog tracks along the muddy fringes of water-bodies. Frogs generally hop, leaving symmetrical groups of prints in which two large, long-toed hind prints fall outside of two smaller fore prints. However, the warm climate of these three states supports too many frog species to make positive track identifications. Although large prints indicate large species, both small species and juveniles of larger species make small prints.

Larger, heavier frogs may leave tracks in soft mud by ponds and woodland paths. The widespread Green Frog (*Rana clamitans*), to 4 inches (10 cm) in length, is often found in wet areas and among fallen trees. The most likely medium-sized frog is the Southern Leopard Frog (*R. spheno-cephala*), which grows to a length of 5 inches (13 cm) and enjoys a variety of wet places. Both display good hopping action, with two small fore prints registering in front of the long-toed hind prints.

An unusually large track is surely from the robust Bullfrog (*R. catesbeiana*), which is found in all regions except for the southernmost parts of peninsular Florida. At up to 8 inches (20 cm) in length, it is the largest frog in North America. It likes to sit along the water's edge, where it waits for prey.

Toads

fore

hind

Straddle
to 2.5 in (6.5 cm)

walking

TOADS

Woodhouse's
Toad

Toad tracks, best
sought along the
muddy fringes
of waterbodies,
can occasionally be
found as unclear trails
in dusty soil. Like frogs (p. 118), toads do hop—especially
when being hassled by overly enthusiastic naturalists—but
in general they walk, leaving rather abstract prints in which
the heels of the hind feet do not register. In mud, the long
toes leave draglines.

Though there are fewer toads than frogs in this region,
Florida boasts both the smallest and largest North American
toads. The tiny Oak Toad (*Bufo quercicus*), up to 1.3 inches
(3.3 cm) long, is found in loose soils of pine forests as well
as in oak scrub. The range of this toad also extends into
southern Georgia and Alabama. This tiny critter would
make an easy meal for the enormous and appropriately
named Giant Toad (*B. marinus*). An introduced species, it
remains confined to southern Florida. It can reach over
9 inches (23 cm) in length, and it will leave the best tracks.

Medium-sized toads include—in all three states—the
Southern Toad (*B. terrestris*) and—in Alabama and Georgia
only—Woodhouse's Toad (*B. woodhousii*). Both may grow
to 5 inches (13 cm) long and prefer loose and sandy soils.
Also resident in loose soils is the Eastern Spadefoot
(*Scaphiopus holbrookii*). Found in all three states, this swift
burrower can grow to just over 3 inches (7.5 cm) in length.

Salamanders

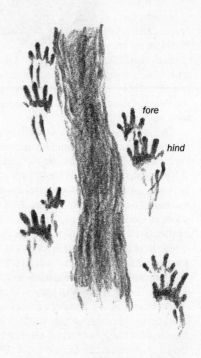

fore

hind

Straddle
to 4 inches (10 cm)

walking

SALAMANDERS & NEWTS

Eastern Tiger Salamander

There are a wealth of salamanders and newts in the region, especially in the mountains of northern Georgia and Alabama. These long, slender lizard-like amphibians are at home in moist or wet areas, such as under logs near waterbodies and in moist woodlands. Look for evidence in soft mud, though good tracks are rare.

Among the more abundant and widespread species is the Eastern Newt (*Notophthalmus viridescens*), which can be up to 5.5 inches (14 cm) long. These newts may leave small trails in mud as they emerge from ponds after rain.

Many species of salamanders are in the region, and they vary enormously in size. Except for in central and southern Florida, the undisputed king of the salamander world has to be the magnificent Eastern Tiger Salamander (*Ambystoma tigrinum*), which can grow up to 14 inches (35 cm) long and comes in such a diversity of colors and patterns that it defies description. Its tracks may have a straddle of up to 4 inches (10 cm).

In general, the fore print shows four toes, and the larger hind print shows five. However, print detail is often blurred by the swinging of the long tail or by a dragging belly. Consideration of size, habitat and distribution will help with track identification.

Lizards

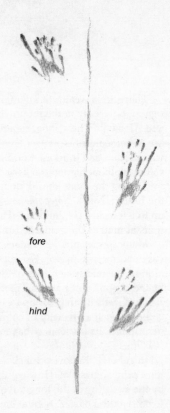

fore

hind

Straddle
to 3 inches (7.5 cm)

walking

LIZARDS

Fence Lizard

The variety of terrain and the warm climate of these states, especially Florida, are perfectly suited for many reptile species. This diversity makes for difficult identification of tracks, which may be blurred by a dragging belly or a thick tail. A lizard's body is similar to a salamander's, as are its tracks, but its toe marks are longer. Also, lizards prefer drier environments than do salamanders.

The Racerunner (*Cnemidophorus sexlineatus*), up to 11 inches (27 cm) long, lives in grassy areas and open woodlands. When cold, it burrows into the soil. The smaller Fence Lizard (*Sceloporus undulatus*), up to 8 inches (20 cm) long, is absent from southernmost Florida but can be seen on rotting logs and grassy dunes elsewhere. The legless Island Glass Lizard (*Ophisaurus compressus*), up to 24 inches (60 cm) long, leaves snakelike (p. 132) tracks on sandy beaches.

Skinks are lizards with an attractive smooth sheen that makes them appear very sleek and swift. One of the skinks most likely to be found throughout the three states is the Ground Skink (*Scincella lateralis*). It lives anywhere from moist forests to partially wooded grasslands, and it can reach up to 5 inches (13 cm) long. The Southeastern Five-lined Skink (*Eumeces inexpectatus*), up to 8.5 inches (22 cm) long, is also likely in similar habitats and in drier areas.

fore

hind

ALLIGATORS, CROCODILES & CAIMANS

American Alligator

The American Alligator (*Alligator mississippiensis*) has a considerable reputation, for obvious reasons. This inhabitant of marshes, ponds and rivers—including brackish water—in Florida and southern regions of Alabama and Georgia is North America's largest reptile: it can grow to the impressive length of over 19 feet (5.8 m).

The fore print shows five toes, and the hind print shows four (webbing may be evident between them). However, other signs are more obvious than the tracks. This relic from the past frequently basks on riverbanks, but it can quickly slide into the water, so look for crushed vegetation and riverside mudslides. If you come across a mound of broken vegetation and mud up to 7 feet (2.1 m) in diameter, it is probably an alligator's nest. Err on the side of caution, because the female is a caring parent and will defend her nest and young. Do not be too tempted to look at any tracks close to the water!

In southernmost peninsular Florida, similar tracks may be from the American Crocodile (*Crocodylus acutus*), which is still hanging on after years of persecution and habitat loss. The much smaller Spectacled Caiman (*Caiman crocodilus*), introduced from South America, is also restricted to the far south.

Freshwater Turtles & Tortoises

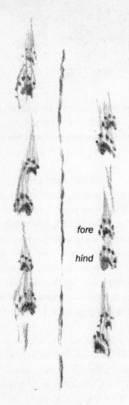

fore

hind

typical turtle walking

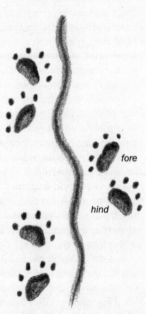

fore

hind

Common Snapping Turtle walking

FRESHWATER TURTLES & TORTOISES

Swamps, lakes and marshes are plentiful in all three states, and so are turtles. They will happily slip into the murky depths to avoid detection, emerging to feed or to bask in the sunshine. Look for their distinctive tracks alongside waterbodies and in moist areas.

Eastern Box Turtle

Some turtles, such as the small Common Musk Turtle (*Sternotherus odoratus*), rarely leave the water. The aquatic Common Snapping Turtle (*Chelydra serpentina*), up to 19 inches (48 cm) long, may leave tracks at pond margins. However, the widespread Eastern Box Turtle (*Terrapene carolina*), up to 8.5 inches (22 cm) long, is a terrestrial species. When the weather is hot and dry, it may leave tracks in damp areas. The Gopher Tortoise (*Gopherus polyphemus*), up to 15 inches (38 cm) long, is a prolific excavator in the sandy soils of southern Georgia and Alabama and most of Florida. Many other animals make their homes in its extensive burrows.

Because of its large shell and short legs, a turtle or tortoise makes a trail that has a straddle about half of its body length. Although longer-legged turtles can raise their shells off the ground, short-legged species may let their shells drag, which is shown in their tracks. The tail may leave a straight dragline in the mud. On firmer surfaces, look for distinct claw marks.

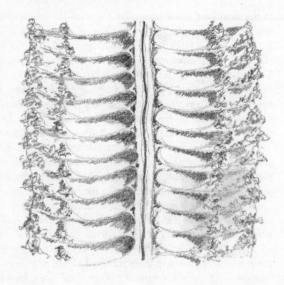

walking

MARINE TURTLES

Hawksbill

Graceful drifters in the great oceans, marine turtles live almost entirely in water. Once a year the females haul themselves onto soft, sandy beaches to lay eggs. Their beautiful tracks are a thrill to see. The four paddle-like limbs make clear dents in the sand, but the huge shell may mostly obliterate the fore prints, and the little tail will leave a dragline.

Loggerhead (*Caretta caretta*) tracks are most likely to be found. Female Loggerheads, up to 4 feet (1.2 m) long, come ashore any time from May to August, preferring wide, sloping beaches. Less commonly encountered is the smaller Hawksbill (*Eretmochelys imbricata*), a rare nester on Florida beaches that grows to only 3 feet (90 cm) long. The beautiful Green Turtle (*Chelonia mydas*), up to 5 feet (1.5 m) in length, occasionally nests on Florida's southeastern coast. The huge Leatherback (*Dermochelys coriacea*) may be found nesting along Florida's east and west coasts. A sight to behold, it can leave phenomenal tracks. It can reach 7 feet (2.1 m) in length and weigh as much as 1600 pounds (730 kg).

Unfortunately, turtles have been heavily hunted in the past. Even today, poachers follow the too obvious turtle tracks to nesting sites and steal the eggs, sometimes even before the female has finished laying them. Consequently, turtle tracks are a rarer sight than they should be.

Snakes

slithering

SNAKES

Common Garter Snake

   Many species of snakes live throughout all three states. Their similar tracks do not give much aid with identification, but consideration of habitat and range may help. A snake track is just a gentle meander, making it very challenging even to establish direction of travel. A wide trail with strong side looping indicates quick movement, whereas narrow and uneven trails result from movement at lower speeds.

   Throughout the region, often close to wet or moist areas, the most widespread, frequently encountered snake is the Common Garter Snake (*Thamnophis sirtalis*). Harmless, it can reach 4.3 feet (1.3 m) in length. Also widespread is the similarly sized Eastern Hognose Snake (*Heterodon platirhinos*), which prefers sandy soils. Several species of watersnakes (*Nerodia* spp.) may leave tracks in mud adjacent to waterbodies. They rarely exceed 5 feet (1.5 m) in length. The longer Mud Snake (*Farancia abacura*), up to 7 feet (2.1 m) long, favors wet areas in southern Georgia and Alabama and all of Florida. The Eastern Indigo Snake (*Drymarchon couperi*), up to 8.6 feet (2.6 m) long, is North America's largest snake and makes very impressive tracks. Its range extends through most of Florida and into Georgia's coastal plains.

TRACK PATTERNS & PRINTS

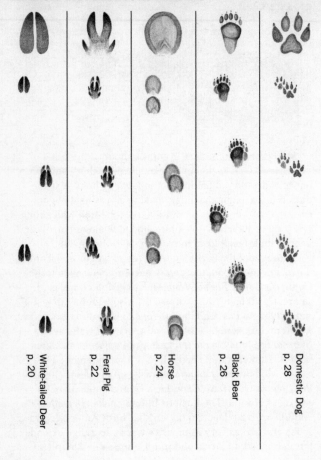

White-tailed Deer
p. 20

Feral Pig
p. 22

Horse
p. 24

Black Bear
p. 26

Domestic Dog
p. 28

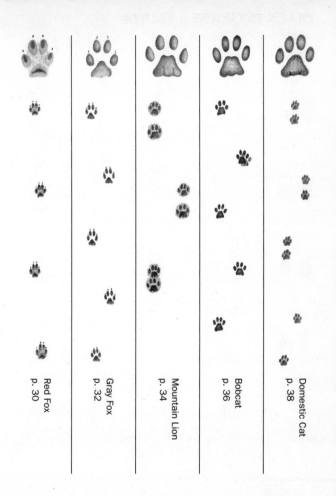

Red Fox
p. 30

Gray Fox
p. 32

Mountain Lion
p. 34

Bobcat
p. 36

Domestic Cat
p. 38

TRACK PATTERNS & PRINTS

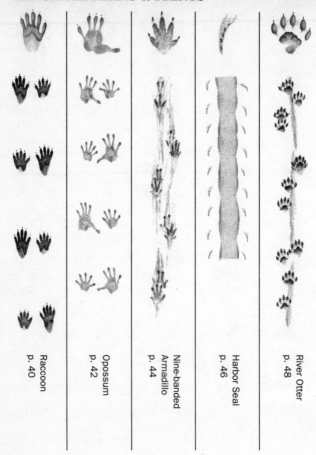

Raccoon
p. 40

Opossum
p. 42

Nine-banded
Armadillo
p. 44

Harbor Seal
p. 46

River Otter
p. 48

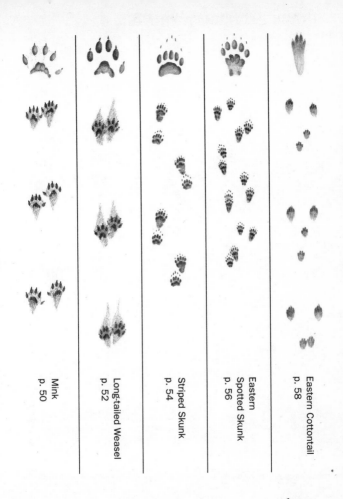

Mink
p. 50

Long-tailed Weasel
p. 52

Striped Skunk
p. 54

Eastern
Spotted Skunk
p. 56

Eastern Cottontail
p. 58

137

TRACK PATTERNS & PRINTS

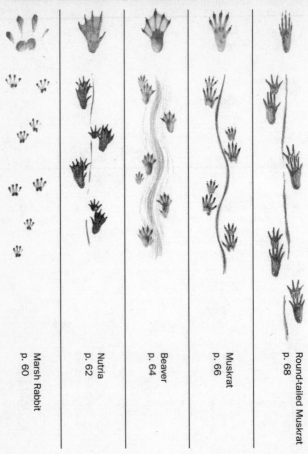

Marsh Rabbit
p. 60

Nutria
p. 62

Beaver
p. 64

Muskrat
p. 66

Round-tailed Muskrat
p. 68

138

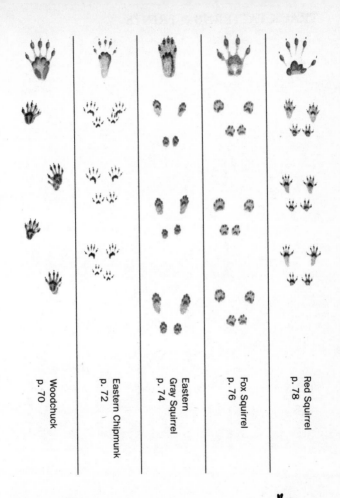

Woodchuck
p. 70

Eastern Chipmunk
p. 72

Eastern
Gray Squirrel
p. 74

Fox Squirrel
p. 76

Red Squirrel
p. 78

139

TRACK PATTERNS & PRINTS

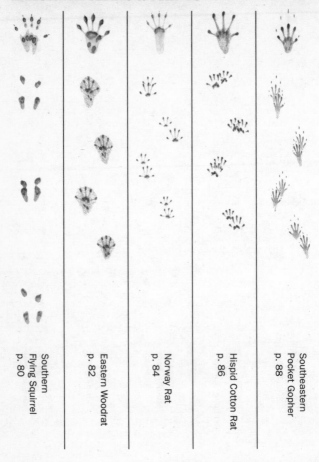

Southern
Flying Squirrel
p. 80

Eastern Woodrat
p. 82

Norway Rat
p. 84

Hispid Cotton Rat
p. 86

Southeastern
Pocket Gopher
p. 88

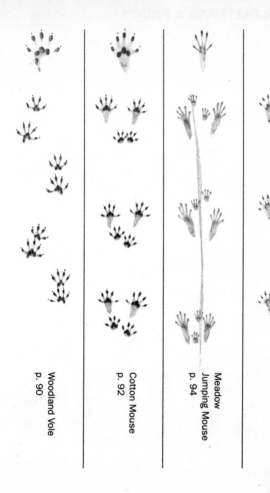

Woodland Vole
p. 90

Cotton Mouse
p. 92

Meadow
Jumping Mouse
p. 94

Southeastern Shrew
p. 96

 141

TRACK PATTERNS & PRINTS

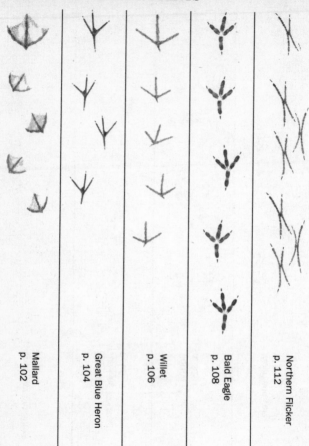

Mallard
p. 102

Great Blue Heron
p. 104

Willet
p. 106

Bald Eagle
p. 108

Northern Flicker
p. 112

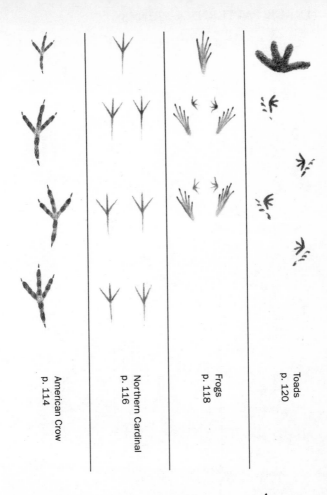

American Crow
p. 114

Northern Cardinal
p. 116

Frogs
p. 118

Toads
p. 120

143

TRACK PATTERNS & PRINTS

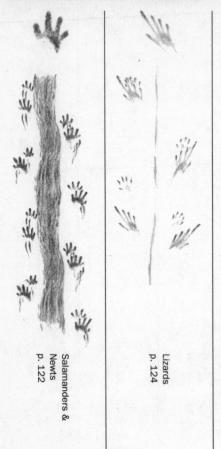

Salamanders &
Newts
p. 122

Lizards
p. 124

Alligators, Crocodiles
& Caimans
p. 126

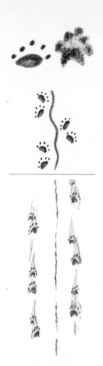

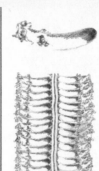

Freshwater Turtles
& Tortoises
p. 128

Marine Turtles
p. 130

Snakes
p. 132

HOOFED PRINTS

White-tailed
Deer

Feral Pig

Horse

inch   cm
0   0

1

2   5

FORE PRINTS

Eastern Spotted Skunk

Striped Skunk

Long-tailed Weasel

Mink

River Otter

inch	cm
0	0
1	
2	5

FORE PRINTS

Gray Fox

Domestic Cat

Red Fox

Bobcat

Domestic Dog

Mountain Lion

HIND PRINTS

Cotton Mouse

Southeastern
Shrew

Woodland
Vole

Hispid
Cotton Rat

Southeastern
Pocket Gopher

Meadow
Jumping Mouse

Norway
Rat

Southern Flying
Squirrel

Eastern
Chipmunk

Eastern
Woodrat

inch   cm
0 —— 0

—— 1

—— 2

1 —— 3

—— 4

2 —— 5

149

HIND PRINTS

Round-tailed
Muskrat

Marsh
Rabbit

Nine-banded
Armadillo

Red
Squirrel

Fox
Squirrel

Eastern
Gray Squirrel

inch cm
0   0

1

2   5

Woodchuck

Muskrat

Eastern
Cottontail

HIND PRINTS

Opossum

Raccoon

Nutria

Beaver

Black Bear

inch cm
0 — 0

1

2 — 5

 151

BIBLIOGRAPHY

Behler, J.L., and F.W. King. 1979. *Field Guide to North American Reptiles and Amphibians.* National Audubon Society. New York: Alfred A. Knopf.

Brown, R., J. Ferguson, M. Lawrence and D. Lees. 1987. *Tracks and Signs of the Birds of Britain and Europe: An Identification Guide.* London: Christopher Helm.

Burt, W.H. 1976. *A Field Guide to the Mammals.* Boston: Houghton Mifflin Company.

Farrand, J., Jr. 1995. *Familiar Animal Tracks of North America.* National Audubon Society Pocket Guide. New York: Alfred A. Knopf.

Forrest, L.R. 1988. *Field Guide to Tracking Animals in Snow.* Harrisburg: Stackpole Books.

Halfpenny, J. 1986. *A Field Guide to Mammal Tracking in North America.* Boulder: Johnson Publishing Company.

Headstrom, R. 1971. *Identifying Animal Tracks.* Toronto: General Publishing Company.

Murie, O.J. 1974. *A Field Guide to Animal Tracks.* The Peterson Field Guide Series. Boston: Houghton Mifflin Company.

Rezendes, P. 1992. *Tracking and the Art of Seeing: How to Read Animal Tracks and Signs.* Vermont: Camden House Publishing.

Stall, C. 1989. *Animal Tracks of the Southeast States.* Seattle: The Mountaineers.

Stokes, D., and L. Stokes. 1986. *A Guide to Animal Tracking and Behaviour.* Toronto: Little, Brown and Company.

Wassink, J.L. 1993. *Mammals of the Central Rockies.* Missoula: Mountain Press Publishing Company.

Whitaker, J.O., Jr. 1996. *National Audubon Society Field Guide to North American Mammals.* New York: Alfred A. Knopf.

INDEX

Page numbers in **boldface** type refer to the primary (illustrated) treatments of animal species and their tracks.

ABOUT THE AUTHOR

Ian Sheldon, an accomplished artist, naturalist and educator, has lived in South Africa, Singapore, Britain and Canada. Caught collecting caterpillars at the age of three, he has been exposed to the beauty and diversity of nature ever since. He was educated at Cambridge University and the University of Alberta. When he is not in the tropics working on conservation projects or immersing himself in the beautiful wilderness, he is sharing his love for nature. Ian enjoys communicating this passion through the visual arts and the written word.